本书系国家社科基金重大研究专项“‘一带一路’沿线国家信息数据库”（项目批准号：17VDL001）成果。

智库 中社

国家智库报告 2020（6）
National Think Tank

人大国发院·国别研究系列

“一带一路”绿色发展报告（2019）

许勤华 著

THE BELT AND ROAD GREEN DEVELOPMENT REPORT 2019

中国社会科学出版社

图书在版编目(CIP)数据

“一带一路”绿色发展报告.2019／许勤华著.—北京：中国社会科学出版社，2020.1

（国家智库报告）

ISBN 978－7－5203－5813－2

Ⅰ.①一… Ⅱ.①许… Ⅲ.①生态环境—环境保护—国际环境合作—研究报告—2019 Ⅳ.①X171.1

中国版本图书馆CIP数据核字（2019）第290476号

出 版 人 赵剑英
责任编辑 喻 苗
责任校对 朱妍洁
责任印制 李寡寡

出 版 中国社会科学出版社
社 址 北京鼓楼西大街甲158号
邮 编 100720
网 址 http://www.csspw.cn
发 行 部 010－84083685
门 市 部 010－84029450
经 销 新华书店及其他书店

印刷装订 北京君升印刷有限公司
版 次 2020年1月第1版
印 次 2020年1月第1次印刷

开 本 787×1092 1/16
印 张 9.5
插 页 2
字 数 95千字
定 价 49.00元

凡购买中国社会科学出版社图书，如有质量问题请与本社营销中心联系调换
电话：010－84083683

“人大国发院·国别研究系列”
编　委　会

总　序

许勤华*

中国人民大学国家发展与战略研究院“一带一路”研究中心集中国人民大学国际关系学院、经济学院、环境学院、财政金融学院、公共管理学院、商学院、社会与人口学院、哲学院、外国语学院和重阳金融研究院的相关人文社科优势学科团队，由许勤华教授、陈甬军教授、王义桅教授、王文教授、戴稳胜教授和王宇洁教授六位首席专家领衔，与中心其他成员共二十位研究员一起，组成了中国人民大学国家高端智库领导下的全校“一带一路”研究的整合平台和跨学科研究团队。

团队围绕“一带一路”建设与中国国家发展、“一带一路”倡议对接沿线国家发展战略、“一带一

* 许勤华，项目执行组长，中国人民大学国际关系学院教授，中国人民大学国家发展与战略研究院副院长、“一带一路”研究中心主任。

路”倡议与新型全球化、“一带一路”倡议关键建设领域四大议题（基础设施投资、文明互鉴、绿色发展、风险治理、区域整合）展开研究。致力于构建“一带一路”沿线国家信息数据库，并在大数据基础上，深入分析沿线国家政治、经济、社会和环境变化，推出“一带一路”智库丛书年度国别系列，为促进“一带一路”建设夯实理论基础、提供政策制定的智力支撑。国别报告对“一带一路”沿线关键合作的64个对象国进行分类研究，规划为文化系列、安全系列和金融系列三类。

习近平主席倡导国与国之间的文明互鉴，强调了文化共融是国际合作成败的基础，深入了解合作国家的安全形势是保障双方合作顺利的前提，资金渠道的畅通是实现“一带一路”建设共商、共建、共享的关键。鉴于目前中国面临世界百年未有之大变局，“一带一路”倡议面临着巨大的机遇与挑战，因此我们首先完成国别研究的安全系列，希冀为“一带一路”合作保驾护航。在国家社科基金重大项目“‘一带一路’沿线国家信息数据库”（项目组长为刘元春教授）完成后，数据库将在条件成熟时，尝试以可视化形式在国发院官网呈现。这也是推出国别报告正式出版物的宗旨。国发院积极为国内外各界提供内部政策报告以及产学研界急需的社会公共研究产品，是中国人民大学作为"世界一流大学"为国家社会科学建设贡献的一分力量。

感谢全国哲学社会科学工作办公室的信任，感谢项目其他两个兄弟单位上海社会科学院和兰州大学的协作，三家在"一带一路"建设重大专项国别和数据库项目研究中通力合作、充分交流，举办了各类学术交流活动，体现了在全国哲学社会科学工作办公室领导下一种成功的、新型的、跨研究机构的合作研究形式，中国人民大学能够作为合作研究的三家单位的秘书处单位深感荣幸。

前　　言

围绕沿线国家和地区的实际环境需求提供多样化的解决方案，既能够增进沿线国家和地区人民福祉、为具体合作落实提供便利，也能促进“一带一路”与全球气候治理等可持续发展议程相协调，使其成为全球治理的重要组成部分。

建设“丝绸之路经济带”和“21 世纪海上丝绸之路”（以下简称“一带一路”）是新时代中国支持多边主义、促进人类社会可持续发展、推动构建人类命运共同体的重要举措。考虑到“一带一路”沿线国家涉及广泛的新兴发展中国家和能源生产消费大国，其环境与气候条件差异较大且总体比较薄弱，妥善应对经济活动等对生态环境带来的压力、转变粗放型经济是各方共同的迫切需求，将绿色发展理念融入“一带一路”建设、形成平衡经济发展与生态环境保护的增长路径是顶层设计中的重要内容。“一带一路”绿色发展对于“一带一路”的成功实施、世界绿色低碳转型以及全球可持续发展目标的成功实施具有重要意义。“一带一路”绿色发展可以帮助其他发展中国家避免依赖传统的高碳增长模式，并寻求中国所展示的更有效和创新的途径。

中国政府一直主张“一带一路”绿色发展，这不仅符合国际社会实现绿色低碳可持续发展的愿景，也与中国新时代高质量发展的内在诉求密切相关。中国国家领导人多次强调，中方将践行绿色发展理念，坚定应对气候变化，倡导绿色低碳可持续的生产生活方式，持续加强生态文明建设。在 2017 年 5 月举行的“一带一路”国际合作高峰论坛上，中国国家主席习近平在开幕式演讲中表示，倡导建立“一带一路”绿色发展国际联盟，促进沿线国家落实联合国《2030 年可持续发展议程》。中国四部委也于 2017 年 5 月联合发布了《关于推进绿色“一带一路”建设的指导意见》。

世界银行将绿色发展定义为“一种将增长与对资源利用、碳排放和环境损害的依赖脱钩，通过创造新的绿色产品市场、技术、投资与改变消费与节约行为来促进增长的发展模式”。但是，综合的绿色发展概念应包含发展过程和结果两个方面。因此，我们将中国传统哲学中的“天人合一”、马克思主义的自然辩证法和当代可持续发展理论的概念结合，在“一带一路”的背景下诠释绿色发展的概念，如图1所示。绿色发展包括一个与“绿色增长”和“可持续发展”密切相关的发展过程，以及与“绿色经济”密切相关的发展阶段。

图1 “一带一路”背景下的绿色发展概念

资料来源：UNDR，2002。

根据“一带一路”背景下的绿色发展概念，中国人民大学建立了一个涵盖国际关系、公共管理、能源治理、环境经济学和气候变化问题的多学科研究团队，基于现有的联合国、世界银行、国际能源署等权威机构的统计数据，对“一带一路”绿色发展指数（Green

Development Index）进行研究，旨在提出相对具体的指标体系，以综合评估“一带一路”国家的绿色发展水平，通过比较分析等找出差距和造成差距的主要影响因素。同时，考虑到该评估方法涉及绿色资产存量、绿色技术创新与绿色发展结果等领域内容，其研究成果也能够用于分析、识别影响绿色“一带一路”建设的关键因素，可服务于确定有助于提高“一带一路”国家绿色发展水平的关键技术、措施等，从而为推动绿色“一带一路”国际合作、切实提高沿线国家或地区绿色低碳能力等相关工作指明方向，支撑形成符合各方诉求的绿色解决方案。

本报告的其余部分将介绍绿色发展指数的设计、数据来源、加权方法、研究结果、政策含义、结论和进一步工作的建议。

摘要：“一带一路”是中国支持多边主义、推动形成合作共赢新型国际关系的全球倡议，也是促进人类社会可持续发展、构建人类命运共同体的重要举措。“一带一路”沿线国家涉及广泛的新兴发展中国家和能源生产消费大国，部分地区环境与气候条件较为薄弱，传统粗放型经济增长模式为其生态环境带来压力，在国际社会致力于推动绿色低碳转型的背景下，各方具有实现经济与环境协调发展的强烈动力与愿望。因此，在“一带一路”建设中倡导绿色发展理念、围绕沿线国家和地区的实际环境需求提供多样化的解决方案，既能够增进沿线国家和地区人民福祉、为具体合作落实提供便利，也能促进“一带一路”与全球气候治理等可持续发展议程相协调，使其成为全球治理的重要组成部分。

本报告利用横跨国际关系、公共管理、能源治理、环境与气候变化经济学的多学科优势，基于现有的联合国、世界银行、国际能源署等权威机构的统计数据，首创了包含20个核心指标的用于综合评估“一带一路”国家绿色发展水平的指标体系。基于指标体系的科学赋权，该小组计算出综合性的“一带一路”绿色发展指数（GDI），定量衡量“一带一路”国家绿色资产存量、绿色技术创新与绿色发展结果三个维度的发展水平，归纳国家间主要的差距及原因，分析、识别在绿色“一带一路”建设发挥关键作用的技术与措施，通过对沿线国家绿色发展情况的梳理比较，把握中国在该进程中的定位，从而为“一带一路”国家合作推进绿色发展指明方向。

本报告有以下发现。

第一，“一带一路”国家绿色发展整体水平有明显提升，但与经合组织国家相比仍然有明显差距。“一带一路”国家的绿色发展指数平均值，由2006年的52.0增长为2015年的54.9，增长率为5.6%，同期经合组织国家的绿色发展指数平均值，由2006年的65.6增长为2015年的68.4，增长率为4.2%。“一带一路”国家的绿色发展指数分值提升速度快于经合组织国家，但是由于基数较低，绝对值差距未见明显缩小。在绿色发展方面，“一带一路”国家未来仍然具有很大的提升潜力。

第二，“一带一路”国家的绿色发展水平与经合组织国家相比，差异性更大，分化明显。2015年“一带一路”国家之间绿色发展指数分值最大差为28.0，而经合组织国家之间最大差仅为22.1。同期，俄罗斯、阿尔巴尼亚、中国、斯洛文尼亚、希腊、捷克、克罗地亚、吉尔吉斯斯坦、波兰、马来西亚等是“一带一路”框架下在绿色低碳领域表现较好的国家，他们的绿色发展得分已经接近经合组织国家平均水平。

2006—2015 年，绿色发展指数分值提升最快的 10 个国家分别是沙特阿拉伯、塞浦路斯、中国、柬埔寨、立陶宛、缅甸、保加利亚、爱沙尼亚、波兰和马来西亚。分区域看，2015 年中东欧国家绿色发展指数平均值最高，达到 60.4，其次是中亚，均值为 55.6。

第三，“一带一路”绿色发展指数由自然资产、绿色技术与发展成果三个维度的分指数构成。2015 年，自然资产分指数排名前十的国家分别是马来西亚、斯洛文尼亚、俄罗斯、爱沙尼亚、拉脱维亚、印度尼西亚、柬埔寨、文莱、克罗地亚和斯洛伐克。绿色技术分指数排名前十的国家分别是中国、阿尔巴尼亚、吉尔吉斯斯坦、土耳其、俄罗斯、印度、土库曼斯坦、卡塔尔、希腊和塔吉克斯坦。发展成果分指数排名前十的国家分别是新加坡、爱沙尼亚、文莱、斯洛文尼亚、沙特阿拉伯、捷克、塞浦路斯、哈萨克斯坦、希腊和卡塔尔。三个分指数中，“一带一路”国家与经合组织国家在绿色技术分指数平均得分上的差距最大，但“一带一路”国家的绿色发展指数分值提升也主要来自绿色技术分指数的增长。2006—2015 年，绿色发展指数增长最快的 10 个国家，其绿色发展指数的增长由绿色技术分指数提升带来的贡献率平均达到 72.0%。这表明，绿色技术的推广应用，既是“一带一路”国家绿色发展的现实薄弱点，又是具有最大提升潜力的重点领域。加强“一带一路”国家的技术合作，加速绿色技术与产品的推广与应用，对于推动“一带一路”绿色发展具有重要意义。

第四，绿色发展指数与人均GDP有较强的相关性，二者之间呈现近似倒“U”形曲线关系。与经合组织国家相比，“一带一路”国家主要位于曲线的左端，在这一区间绿色发展指数与人均GDP的相关性更强，这表明，在人均GDP较低阶段，经济增长与绿色发展水平提升在方向上具有一致性。因此，“一带一路”国家基于其所处的发展阶段和特点，发展经济仍然是首要任务。同时，发展经济也可以跟绿色发展同步提升。但是在经济发展过程中，“一带一路”国家亟须跨越传统的“先污染，后治理”发展路径，依靠绿色技术上的后发优势实现向绿色发展模式的转型。

第五，基于绿色发展指数的评估，中国在2006—2015年这10年在绿色发展领域取得了积极进展，经评估成为同期指数分值增长最快的国家之一。中国2015年的绿色发展指数分值在“一带一路”国家中排名第三，已经接近经合组织国家2006年的平均水平，其中绿色技术分指数为71.1，在“一带一路”国家中排名第一，是这10年间中国绿色发展指数分值大幅跃升的重要贡献因素。在该阶段，中国高度重视提高能源效率与资源集约工作，大力支持可再生能源等低碳技术研发、示范和推广应用，持续加大循环经济、节能节水等工作力度，推动相关领域技术创新。这一成就也表明，作为最大的发展中国家和新兴经济体，中国通过加强生态文明建设、积极应对气候变化，正逐步开辟出符合发展中国家国情与需要的绿色增长路径，相关经验能够为沿线国家或地区提供可借鉴、可复制、可操作性的示范案例。这也预示着，“一带一路”覆盖的大量的发展中国家在绿色低碳领域基础设施建设、技术研发等方面具有广阔的合作空间。

关键词：“一带一路”；绿色发展指数；中国

Abstract: The Belt and Road Initiative (BRI) is a global initiative for supporting multilateralism and building a new type of international relations with win-win cooperation. It is also regard as actions towards sustainable development and a community of shared future for mankind. BRI covers plenty of emerging economies and major energy powers. Meanwhile, some regions of these areas are climate vulnerable and suffered from "extensive" mode of economic growth. In the context of green low-carbon transition, all parties could reach consensus on coordinating development between economy and environment. In this regard, integrating the green development concept into the establishment of the BRI could improve people's wellbeing and facilitate cooperation among relevant parties. It'll also bridge the gap between the BRI and the 2030 Agenda for sustainable development which could contribute to the global governance.

Based on authoritative statistics from United Nations, World Bank, International Energy Agency and other relevant organizations, we take advantage of inter-disciplinary for deepening our research and create the Green Development Index (GDI) System which covers roughly 20 indicators related to the green development standard. Under such condition, our team has calculated the specific figures of the countries along the Belt and Road which could reflect their capacities of natural assets, green technology and development achievements. Afterwards, we analyze main reasons on the different performances among relevant countries and figure out the key technology or measures related to the establishment of green Belt and Road. Based on these analyses and comparisons, our research tries to figure out China's position in this process and provide suggestions for facilitating cooperation among relevant countries in green development field.

Our main points are as follow: firstly, BRI countries have made great progress in green development field in the past 10 years. However, there are still huge gaps between their performance and OECD member states'. From 2006 to 2015, the average GDI number of countries covered by BRI has increased from 52.0 to 54.9 which increased by 5.6%. Meanwhile, the average GDI number of OECD member states has increased from 65.6 to 69.4 which increased by 4.2%. It seems BRI countries have achieved better performance according to the growth rate, while the gap of numerical difference hasn't showed clear sign of narrowing. It also reflects that there is still great potential for BRI countries to strengthen cooperation.

Secondly, the numerical difference among BRI countries is larger than the difference among OECD according to calculation mentioned above. The hugest gap between the Belt and Road countries was 28.0, while that of the latter was only 22.1. In 2015, regarding GDI, the top ten countries of the Belt and Road were Russia, Albania, China, Slovenia, Greece, Czech republic, Croatia, Kyrgyzstan, Poland, Malaysia, and their green development indexes were close to the average level of the OECD countries. As for the growing speed during 2006 to 2015, the fastest ten countries in GDI were Saudi Arabia, Cyprus, China, Cambodia, Lithuania, Myanmar, Bulgaria, Estonia, Poland and Malaysia. In consideration of regional green development, the Central and Eastern Europe ranks first and scores 60.4 in 2015, while Central Asia ranks second and scores 55.6.

Thirdly, the GDI is a comprehensive index to evaluate the natural assets, green technology and development achievements of various BRI countries. In 2015, Malaysia, Slovenia, Russia, Estonia, Latvia, Indonesia, Cambodia, Brunei, Croatia and Slovakia rank top 10 among BRI countries in natural assets area. China, Albania, Kyrghyzstan, turkey, Russia, India, Turkmenistan, Qatar, Greece and

Tajikistan rank top 10 in green technology area. Singapore, Estonia, Brunei, Slovenia, Saudi Arabia, Czech, Cyprus, Kazakhstan, Greece and Qatar rank top 10 in development achievements area. Among these three sub-indexes, the largest difference between BRI countries and OECD appears in green technology area which also makes great contribution to the GDI score improvement of BRI countries. During the period of 2006 to 2015, among the top ten GDI increasing countries, it was the increase of green technologies index that contributed to the growth of the whole, reaching to an average of 72.0%. This indicates that the promotion and application of green technologies are not only the weak point of the green development of the participating countries, but also a key area which has the greatest potential for improvement. It is of great importance that we should cooperate more in technology and accelerate the promotion and application of green technologies and products, which can improve the green development of the partner countries.

Fourthly, GDI has close relation with GDP per capita which can be approximately described as inverted U-shaped curve. Compared with OECD countries, the Belt and Road countries are mainly located on the left bottom of the curve, in which interval the GDI is more correlated with GDP per capita. It shows that in the lower stage of GDP per capita, there is a consistent trend in direction between the economy development and green development. Therefore, based on the stage of development and characteristics of the BRI countries, the top priority of the partner countries is still to develop the economy. Meanwhile, the green development can be simultaneously upgraded with the economy development. However, in the process of economy development, BRI countries need to abandon the traditional development path and should transform to the green development way with the advantage of backwardness of the green technologies.

Finally, China witnessed a growth in GDI from 2006 to 2015 which is one of the fastest growing countries. The GDI level of China in 2015 ranked third among the BRI countries, which was close to the average level of OECD countries in 2006. Besides, the green technologies sub-index of China is 71.1, ranking first among the Belt and Road countries. The increase in the green technologies sub-index mainly explained for the significant increase in GDI of China , which is due to China's great efforts in research, promotion and innovations of green low-carbon technologies, such as energy efficiency, renewable energy and circular economy. The outcomes prove that, as the largest developing country and emerging economy, China is exploring and opening up a green sustainable development way through strengthening ecological civilization construction, addressing climate change and other relevant methods, which provides models and experiences that can be used for reference for other developing countries. It also means that there is an immense potential for cooperation between China and the Belt and Road countries in green low-carbon fields.

Key Words: The Belt and Road; the Green Development Index (GDI); China

目　　录

第一章 “一带一路”绿色发展指数[*]

第一节 框架

根据定义，“发展”既可指进步的过程，也可指实现某种程度的经济、社会与环境成果。绿色发展亦是如此。在本指标体系中，绿色发展分为两个发展状态和一个发展过程（见图2）。自然资产为未来发展奠定基础或构成限制，是经济的禀赋和初始状态。发展成果可随时间累积，是整体绿色发展的结果和现状。在将自然资产转化为发展成果的过程中，绿色技术在塑造经济和引领发展方向中发挥着重要作用。这三个类别是绿色发展全过程不可或缺的组成部分，因此全部纳入绿色发展指数体系。

---

* 本书为国家社科重大研究专项“推动绿色‘一带一路’建设研究”（18VDL009）与国家社科一般项目“新时代中国能源外交战略研究”（18BGJ024）的阶段性成果。

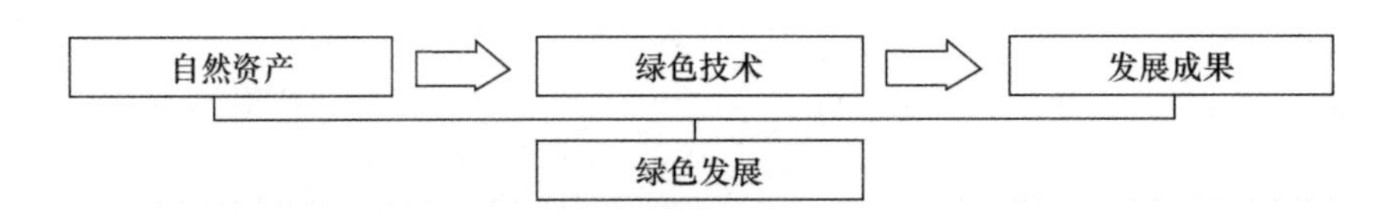

图 2 “一带一路”绿色发展指数框架

第二节 分类及指标

在上述研究框架下，研究组建立了一个支持索引系统的数据库。数据库覆盖了 98 个国家，包括“一带一路”沿线的 65 个国家、中国以及所有经合组织国家。指标体系由 3 个类别、15 个子类别和 20 个指标组成，时间跨度从 2006 年到 2015 年。考虑到小国内部发展与大国权力的溢出效应之间的平衡，总体指标和平均水平指标都包括在内。数据主要来自官方和可靠来源，包括世界银行、联合国开发计划署、联合国环境规划署、国际能源署、国际可再生能源署、联合国教科文组织、耶鲁大学和联合国大学等（见表 1）。

表 1　　“一带一路”国家绿色发展指数指标体系

指标分类	指标子类别	指标及单位	评价方向	指标维度	数据来源
自然资产	森林资源	森林覆盖率（%）	正	平均	WB
	生物多样性	生物多样性和栖息地（EPI 标准化得分）	正	平均	EPI
	水资源	人均可再生淡水资源（m^3/人）	正	平均	WB
	自然资源	自然资源租金总额（美元）	正	规模	WB
		自然资源租金总额占 GDP 比重（%）	正	平均	

续表

指标分类	指标子类别	指标及单位	评价方向	指标维度	数据来源
绿色技术	可再生能源发电	可再生能源发电量（GWh）	正	规模	IEA
		可再生能源发电量占总发电量比重（%）	正	平均	UN
	可再生能源装机	可再生能源装机容量（GW）	正	规模	
		可再生能源装机容量占总装机容量的比重（%）	正	平均	
绿色技术	能源效率	人均 GDP 能源消费量（kg toe/美元）	负	平均	WB
	绿色交通	人均交通部门二氧化碳排放量（kg/人）	负	平均	IEA
	绿色建筑	研究和开发投资占 GDP 比重（%）	正	平均	IEA
	技术研发竞争力	研究和开发投资（百万美元）	正	规模	WB
		研究和开发投资占 GDP 比重（%）	正	平均	
发展成果	人类发展指数	人类发展指数	正	平均	UNDP
	不平等	基尼系数	负	平均	UNU
	通电率	可用电人口占总人口比重（%）	正	平均	WB
	碳排放	人均燃料燃烧二氧化碳排放（kg/人）	负	平均	IEA
	PM2.5	年均 PM2.5 浓度（mg/m^3）	负	平均	EPI
		PM2.5 平均暴露度（EPI 标准化得分）	正	平均	

注：WB 为世界银行；EPI 为环境治理指数（耶鲁大学地球科学信息网络环境法与环境政策中心，联合哥伦比亚大学及世界经济论坛合作）；IEA 为国际能源署；UN 为联合国；经合组织为经济合作与发展组织；UNDP 为联合国开发计划署；UNU 为联合国大学。

第二章　数据处理及指标赋权

第一节　数据标准化

“一带一路”国家绿色发展指数是用于评估“一带一路”国家自然资产、绿色技术和发展成就等绿色发展相关要素实际表现的综合指标。它由三个子指标组成，每个子指标由若干定量指标合成。由于正向指标（值越大，绿色发展水平越高）和负向指标（值越小，绿色发展水平越高）的存在，以及指标之间的数量和单位差异较大，在计算子指标之前，研究组对这些指标进行了标准化以便进行各国之间指数的横向比较和一国不同时期指数的纵向比较，在比较各国绿色发展相对水平的同时，也可以考察同一国家绿色发展水平的历史发展过程。因此，原始数据的标准化是必不可少的。

数据标准化根据数据特点采取了不同的方法。对于数量级较低的指数，直接进行无量纲化处理，正向指标无量纲化处理公式为：

$$Z_i = \frac{X_i - min\ (X_i)}{max\ (X_i) - min\ (X_i)} \times 100\% \qquad (1)$$

$$Z_i = \frac{max\ (X_i) - X_i}{\max(X_i) - min\ (X_i)} \times 100\% \qquad (2)$$

其中，X_i 表示第 i 个指标各年份的取值，$min\ (X_i)$ 表示第 i 个指标基数年的最小值，$max\ (X_i)$ 表示第 i 个指标基数年的最大值。

对于数量级较高的指数，对指数值进行取对后再进行无量纲化操作：

$$Z_i = \frac{Ln\ X_i - min\ (Ln\ X_i)}{max(Ln\ X_i) - min\ (Ln\ X_i)} \times 100\% \qquad (3)$$

$$Z_i = \frac{max(Ln\ X_i) - Ln\ X_i}{max(Ln\ X_i) - min\ (Ln\ X_i)} \times 100\% \qquad (4)$$

第二节　指标权重的确定

权重决定了指标体系中每个指标的相对重要性，从而决定了被评估国家的最终排名。权重的作用是将各种指标组合成一个综合统一的绿色发展指数。虽然我们收集了大量的数据，并确保绿色发展指数系统（见表1）中列出的所有指标都与绿色发展相关，但这些指标与未观测到的绿色发展指数之间的相关性可能非常不同。不同的权重倾向于强调绿色发展的不同方面，并且可能没有最佳的权重系统。

在指标体系中，有多种赋权方法：一是比较主观的方法，如德尔菲法和层次分析法，二是比较客观的方法，如主成分分析法和因子分析法。此外，当很难区分两个指标时，为指标分配相等的权重的做法也很常见，例如人类发展指数所采用的权重分配体系。

本报告利用因子分析的统计技术提出了表 1 中列出的 20 个指标的权重。在使用因子分析时，限制了分析，从而提取出了指标体系 20 个指标中最重要的三个共同因素。基于这些指标与三个共同因素的相关性，这三个共同因素分别被称为发展、技术和环境，有趣的是，它们与绿色发展定义的三个关键组成部分非常相似。这首先使我们对绿色发展的理论定义有了更多的信心，同时也体现出了因子分析的作用。

20 个绿色发展指数指标的最终权重如表 2 所示。根据研究结果，虽然发展（结果）是提取出的因素中最重要的一个，但权重最高的是（绿色）技术，权重为 56. 3。这是因为绿色技术类别中的若干指标也与发展共同因素高度相关，例如建筑物中的 CO_2 排放以及研发支出，因此这些指标将从发展共同因素中获得一些权重。

在绿色技术范畴内，技术研发竞争力、可再生能源发电和可再生能源装机是权重最高的三大要素。在发展成果类别中，HDI、CO_2 排放和获得电力相对更为重要。最后，在自然资产类别中，水资源似乎比其他资产更重要，突显了“一带一路”国家淡水资源的稀缺性。

表 2　　“一带一路”国家绿色发展指数指标权重

指标分类（权重）	指标子类别（权重）	指标及单位	最终权重
自然资产（17.3）	森林资源（3.6）	森林覆盖率（占土地面积的百分比）	3.6
	生物多样性（3.4）	生物多样性和栖息地（EPI 指标，已标准化）	3.4
	水资源（5.5）	人均可再生淡水资源（m^3/人）	5.5
	自然资源（4.9）	自然资源租金总额（百万美元）	2.6
		自然资源租金总额占 GDP 比重（%）	2.3
绿色技术（56.3）	可再生能源发电（12.3）	可再生能源发电量（GWh）	8.0
		可再生能源发电量占总发电量比重（%）	4.2
	可再生能源装机（11.8）	可再生能源装机容量（GW）	7.4
		可再生能源装机容量占总装机容量的比重（%）	4.4
	能源效率（6.5）	人均 GDP 能源消费量（kg toe /美元）	6.5
	绿色交通（5.0）	人均交通部门二氧化碳排放量（kg/人）	5.0
	绿色建筑（5.0）	人均建筑部门二氧化碳排放量（kg/人）	5.0
	技术研发竞争力（15.7）	研究和开发投资（百万美元）	8.6
		研究和开发投资占 GDP 比重（%）	7.1
发展成果（26.4）	人类发展指数（9.2）	人类发展指数	9.2
	不平等（1.0）	基尼系数	1.0
	通电率（4.6）	可用电人口占总人口比重（%）	4.6
	碳排放（1.6）	人均燃料燃烧二氧化碳排放（kg/人）	6.5
	PM2.5（5.0）	年均 PM2.5 浓度（mg/m^3）	1.6
		PM2.5 平均暴露度（EPI 标准化得分）	3.4

注：列中的权重总计为 100。

第三节 分类指数和指数的总合成

利用上述提供的权重（W_j），可以计算出绿色发展指数和每个国家和年度的子指数。对于总绿色发展指数：

$$GDI_{it} = \sum_j W_j X_{ijt} \tag{5}$$

其中，i 表示一个国家，t 表示一年，j 表示指标。因此，绿色发展指数GDI_{it} 是 20 个不同指标 X_{ijt} 的加权平均值。类似的，我们可以计算自然资产、绿色技术和发展成果的绿色发展指数（子类别指数）。

第三章　结果计算与分析

第一节　基于绿色发展指数的国家排名

在上述研究方法的基础上，本报告计算了“一带一路”国家绿色发展指数及其子类别指数值。表 3 显示了 2015 年排前 20 位的国家（绿色发展指数及其子类别指数）的结果[①]。如表 3 所示，俄罗斯、阿尔巴尼亚和中国占据前 3 名。值得注意的是，俄罗斯获得第一，主要是因为它在自然资源、绿色技术和发展成果方面的得分相对较高。中国在 2015 年排名第三，这主要是由于绿色技术分数的不断提高。

表 4 列出了经合组织排前 20 位的国家。相比经合组织国家，“一带一路”国家的绿色发展水平差异更大。“一带一路”国家之间绿色发展指数分值最大差

① 表中绿色发展指数和子类别的值都是标准化的。

为28.0，而后者之间最大差只有22.1。基于2015年的绿色发展指数评估结果，俄罗斯、阿尔巴尼亚、中国、斯洛文尼亚、希腊、捷克、克罗地亚、吉尔吉斯斯坦、波兰、马来西亚位列前十，其指数分值已经接近经合组织国家的平均水平。可以看出，与“一带一路”国家前20名相比，经合组织国家的绿色发展指数相对较高。实际上，对“一带一路”国家和经合组织国家绿色发展指数时间序列的比较分析表明，经合组织的年平均指数（2006—2015）为67.4，而“一带一路”国家仅为54.3。

“一带一路”绿色发展指数包括三个维度的子指标——自然资源、绿色技术和发展成果。根据三个子指数得分，排名前十的国家的结果如表5所示。2015年，自然资源指数排名前十的国家是马来西亚、斯洛文尼亚、俄罗斯、爱沙尼亚、拉脱维亚、印度尼西亚、柬埔寨、文莱、克罗地亚和斯洛伐克。绿色技术指数排名前十的国家是中国、阿尔巴尼亚、吉尔吉斯斯坦、土耳其、俄罗斯、印度、土库曼斯坦、卡塔尔、希腊和塔吉克斯坦。在发展成果指数中排名前十的国家是新加坡、爱沙尼亚、文莱、斯洛文尼亚、阿拉伯联合酋长国、捷克、塞浦路斯、哈萨克斯坦、希腊和卡塔尔。

在这三个分指标中，“一带一路”国家与经合组织国家之间差距最大的是绿色技术指数。同样，经合组织国家绿色发展指数的增长主要归因于绿色技术指数的上升。2006—2015 年，全球十大绿色发展指数增长国中，绿色技术指数增长对整体增长贡献最大，平均达到 72.0%。这表明，绿色技术的推广与应用不仅是参与国绿色发展的薄弱环节，也是最具发展潜力的关键领域。在技术上，我们应该与"一带一路"国家进行更多的合作，加快绿色技术和产品的推广与应用，促进合作伙伴国的绿色发展。

表 3　排名前 20 的"一带一路"国家（绿色发展指数 2015 年）

排名	国家	绿色发展指数子类别：自然资源	绿色发展指数子类别：绿色技术	绿色发展指数子类别：发展成果	绿色发展指数
1	俄罗斯	70.9	57.6	81.7	66.3
2	阿尔巴尼亚	53.5	68.2	69.7	66.1
3	中国	60.6	71.1	57.6	65.7
4	斯洛文尼亚	73.6	52.9	84.5	64.8
5	希腊	64.1	54.7	82.4	63.7
6	捷克	62.5	54.0	83.1	63.2
7	克罗地亚	66.5	54.6	77.1	62.6
8	吉尔吉斯斯坦	51.7	64.3	65.5	62.4
9	波兰	63.3	53.8	80.2	62.4
10	马来西亚	75.3	50.5	78.4	62.2
11	拉脱维亚	70.6	49.9	79.7	61.3
12	罗马尼亚	62.2	54.7	74.8	61.3
13	土耳其	45.9	60.3	73.3	61.3
14	斯洛伐克	65.8	50.0	79.7	60.5

续表

排名	国家	绿色发展指数子类别：自然资源	绿色发展指数子类别：绿色技术	绿色发展指数子类别：发展成果	绿色发展指数
15	立陶宛	65.1	48.8	79.3	59.7
16	爱沙尼亚	70.7	43.4	87.0	59.6
17	卡塔尔	38.0	55.0	82.4	59.3
18	格鲁吉亚	61.8	51.8	71.7	58.8
19	保加利亚	61.0	49.4	76.4	58.6
20	以色列	43.6	50.1	82.3	57.4

表4　　排名前20的经合组织国家（绿色发展指数2015年）

排名	国家	绿色发展指数子类别：自然资源	绿色发展指数子类别：绿色技术	绿色发展指数子类别：发展成果	绿色发展指数
1	挪威	69.4	75.9	92.0	79.0
2	瑞典	75.7	73.7	87.3	77.6
3	加拿大	71.5	73.2	91.0	77.6
4	瑞士	63.5	78.1	84.8	77.3
5	奥地利	69.5	75.8	84.9	77.1
6	美国	66.8	71.6	90.3	75.7
7	日本	74.0	70.8	86.6	75.5
8	德国	61.1	74.5	87.2	75.5
9	芬兰	79.7	66.5	89.8	74.9
10	丹麦	57.0	70.2	88.2	72.7
11	澳大利亚	63.3	63.8	93.5	71.5
12	法国	65.7	67.5	83.5	71.4
13	冰岛	59.3	66.7	88.7	71.2
14	意大利	66.0	67.7	80.8	70.8
15	新西兰	73.0	61.4	88.5	70.5
16	西班牙	65.6	65.7	84.0	70.5

续表

排名	国家	绿色发展指数子类别：自然资源	绿色发展指数子类别：绿色技术	绿色发展指数子类别：发展成果	绿色发展指数
17	韩国	65.4	66.5	79.4	69.7
18	葡萄牙	64.1	63.8	81.4	68.5
19	爱尔兰	58.0	62.4	88.2	68.4
20	比利时	58.4	63.8	83.8	68.1

表5　2015年绿色发展指数子类别排名前10的"一带一路"国家

排名	国家	绿色发展指数子类别：自然资源	国家	绿色发展指数子类别：绿色技术	国家	绿色发展指数子类别：发展成果
1	马来西亚	75.29	中国	71.12	新加坡	87.06
2	斯洛文尼亚	73.56	阿尔巴尼亚	68.24	爱沙尼亚	87.05
3	俄罗斯	70.89	吉尔吉斯斯坦	64.32	文莱	86.93
4	爱沙尼亚	70.69	土耳其	60.34	斯洛文尼亚	84.50
5	拉脱维亚	70.55	俄罗斯	57.58	阿拉伯联合酋长国	83.47
6	印度尼西亚	68.71	印度	57.01	捷克	83.11
7	柬埔寨	67.00	土库曼斯坦	56.63	塞浦路斯	83.02
8	文莱	66.71	卡塔尔	54.98	哈萨克斯坦	82.90
9	克罗地亚	66.54	希腊	54.71	希腊	82.43
10	斯洛伐克	65.76	塔吉克斯坦	54.67	卡塔尔	82.38

第二节　"一带一路"国家与经合组织国家对比分析

如图3所示，与经合组织国家相比，"一带一路"国家的绿色发展指数平均值较低，而且分布格局通常较为分散。2015 年，“一带一路”国家的绿色发展指数极值为 28.0，而经合组织国家的绿色发展指数极值为 22.1。此外，不同地域国家群体之间存在明显的绿色发展指数分数差异（见图 4）。2015 年，中东欧的 15 个国家的绿色发展指数平均值为 60，中亚的 5 个国家绿色发展指数平均值为 56。

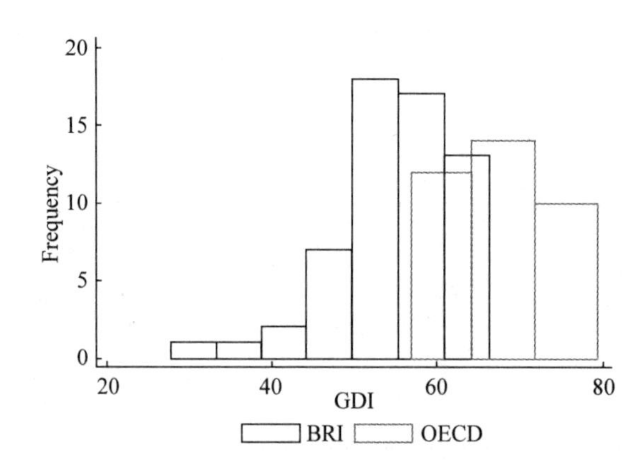

图 3　2015 年“一带一路”国家和经合组织国家绿色发展指数分布

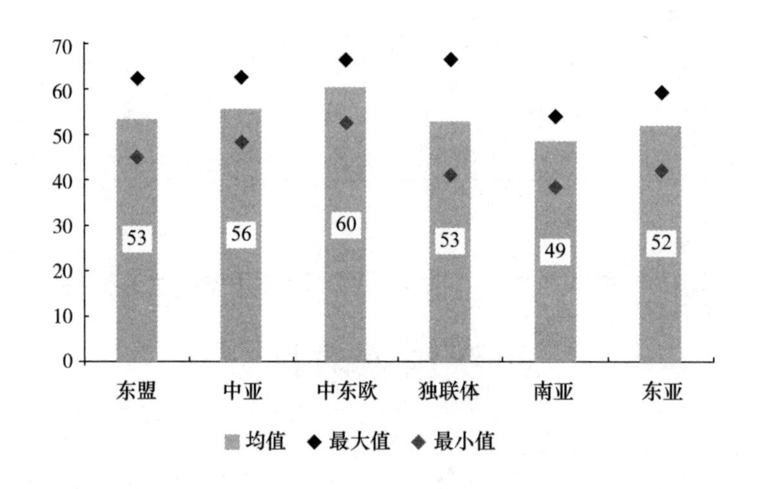

图 4　2015 年不同地理区域国家绿色发展指数

第三节　绿色发展指数变化趋势分析

在分析的时间跨度内，“一带一路”国家在绿色发展方面取得了巨大进步，这些进展得到了指数体系的充分解释。“一带一路”国家的绿色发展指数均值从2006年的52.0上升到了2015年的54.9，增长率为5.6%，高于经合组织国家4.2%的增长率。2006—2015年绿色发展指数增长最快的10个国家是沙特阿拉伯、塞浦路斯、中国、柬埔寨、立陶宛、缅甸、保加利亚、爱沙尼亚、波兰和马来西亚，绿色发展指数得分平均增长7.6。中国在绿色发展指数增长最快的10个国家中排名第三。这一增长主要得益于“一带一路”国家绿色技术的发展，绿色技术为绿色发展指数增长最快的10个国家平均贡献了72.0%的指数增加值。

此外，绿色发展指数与人均GDP之间存在很强的相关性，可以近似描述为倒“U”形曲线。与经合组织国家相比，“一带一路”国家主要位于曲线的左下方，其中绿色发展指数与人均GDP的相关性更大。结果表明，在人均GDP较低的阶段，经济发展与绿色发展之间存在着一致的趋势。因此，根据“一带一路”国家的发展阶段和特点，大部分国家的首要任务仍然是发展经济。同时，随着经济的发展，绿色发展可以同步升级。但是，在经济发展过程中，“一带一路”沿线国家需要放弃传统的发展路径即“先污染后治理”的路径，应该转向绿色发展道路，依靠绿色技术的优势提升绿色发展水平。

表 6 2006—2015 年绿色发展指数名次提升排名前 10 的“一带一路”国家

国家	绿色发展指数增加值	绿色发展指数提升名次
沙特阿拉伯	11.07	28
塞浦路斯	9.19	16
中国	8.49	10
柬埔寨	8.23	5
立陶宛	6.65	12
缅甸	6.57	12
保加利亚	6.49	10
爱沙尼亚	6.48	9
波兰	6.31	9
马来西亚	6.11	9

第四节 中国绿色发展指数变化趋势

中国的绿色发展指数从 2006 年的 57.2 增长到 2015 年的 65.7，是增长最快的国家之一。2015 年中国

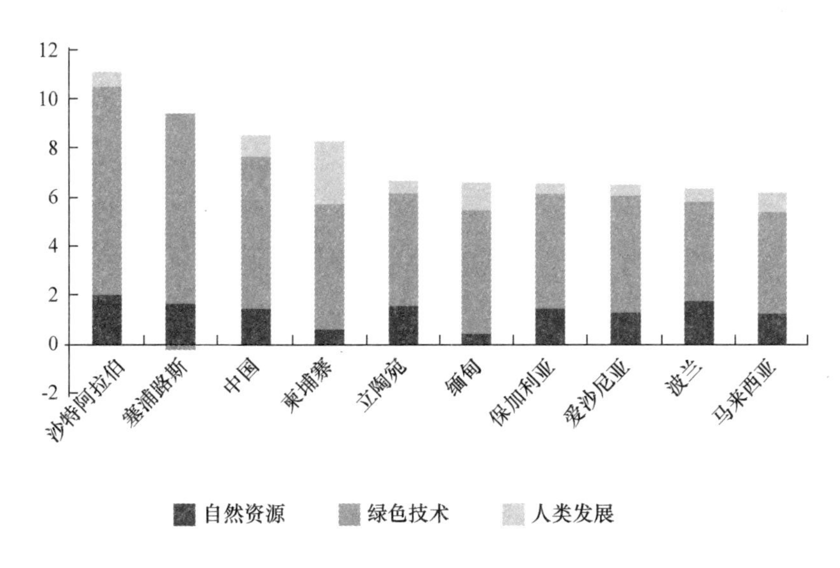

图5 2006—2015年绿色发展指数变化量（子指标）

的绿色发展指数水平在“一带一路”国家中排名第三，接近2006年经合组织国家的平均水平。此外，中国绿色技术指数为71.1，位居“一带一路”国家的绿色技术指数第一位。绿色技术指数的增长主要来源于中国在研究、示范和推广绿色低碳技术（如能效和可再生能源技术）方面的巨大努力。近年来，中国绿色发展指数水平的显著上升反映出中国作为最大的发展中国家，正逐步通过创新发展道路探索适合发展中国家的开放、绿色的可持续发展道路。它提供了可供其他发展中国家参考的模型和经验。在如绿色基础设施建设和标准制定、绿色技术研究、示范和推广的领域，中国与“一带一路”国家之间的合作潜力巨大。

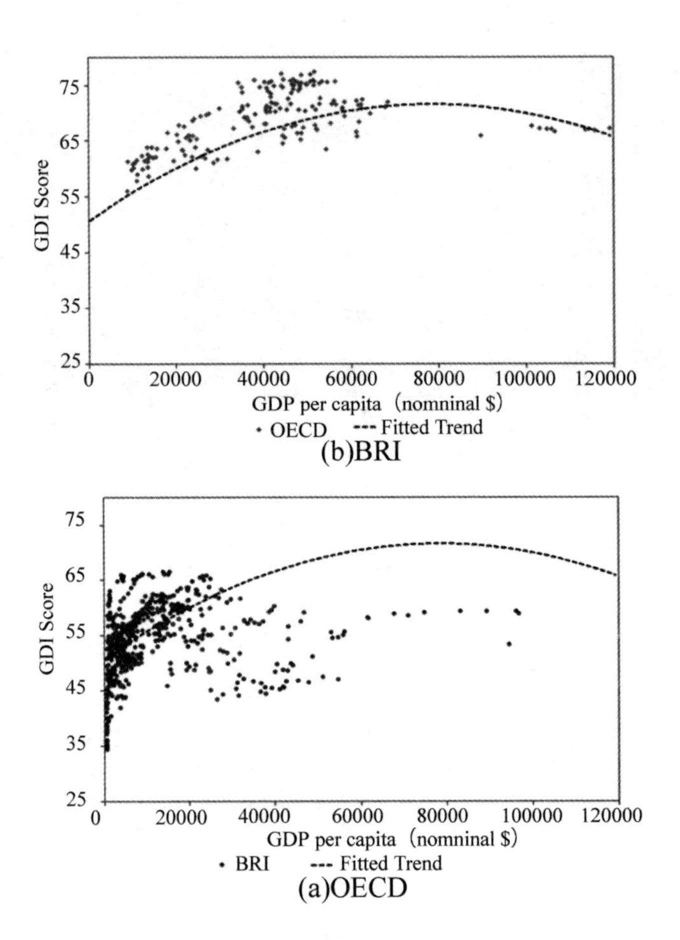

图6　绿色发展指数与人均GDP的库兹涅茨曲线

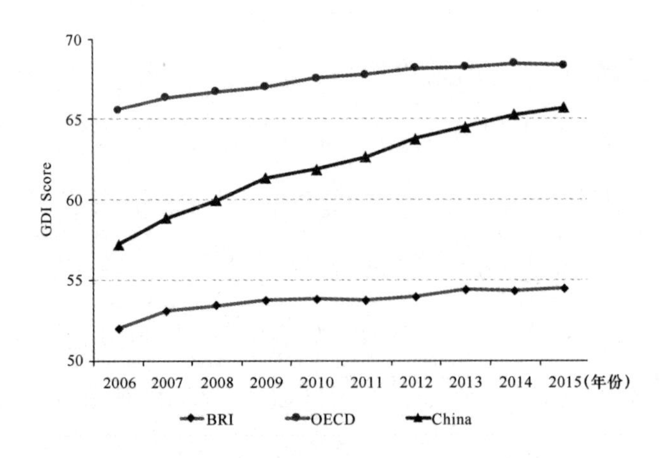

图7　"一带一路"国家和经合组织国家绿色发展指数随时间变化情况

第四章　绿色发展指数分析的政策含义

上述绿色发展指数结果分析表明，绿色发展的表现因国家而异；对于“一带一路”国家来说尤其如此。2015年“一带一路”国家绿色发展指数最高为66，最低仅为28，相差近40。经合组织国家绿色发展指数最高为79，最低为57，仅相差22。因此，与经合组织国家相比，“一带一路”国家不仅面临经济发展的任务，而且总体上也不那么绿色。尽管俄罗斯和阿尔巴尼亚等“一带一路”国家已经接近某些经合组织国家的发展水平，但在过去10年中，它们的发展状况并未发生太大变化。

“一带一路”国家的低起点也为它们提供了巨大的绿色发展潜力。在观察绿色发展指数随时间变化的情况时，沙特阿拉伯、塞浦路斯和中国等国家脱颖而出，这些国家都是从低绿色发展指数分数开始的。有趣的是，它们的改进指向了相同的潜在原因——绿色技术，尽管具体技术略有不同：沙特阿拉伯由于其研发和能源效率而提高，而塞浦路斯和中国由于其可再生能源开发和能源效率而提高。这表明一个国家可以通过其他国家的成功经验学习等多种渠道来攀登绿色发展指数阶梯。因此，将绿色技术纳入绿色发展指数体系对于一个国家实现更好的绿色发展是必要且有成效的。

绿色发展指数得分分析的另一个重要政策含义是绿色发展不等于发展成果，后者包括人均 GDP，或更广泛的人类发展指数。总的来说，绿色发展指数与人均 GDP 之间存在倒“U”形关系，这意味着当人均 GDP 达到一定水平时，绿色发展指数得分就会停止增加。这既适用于经合组织国家，也适用于“一带一路”国家。在“一带一路”国家之中，新加坡、文莱和阿拉伯联合酋长国等国家在发展成果方面表现良好；然而，他们在整体绿色发展指数中得分都很低。为应对全球气候变化挑战，“一带一路”国家应采用更全面的发展概念，并开始接受绿色发展的理念。

中国可以成为“一带一路”国家和经合组织国家之间的有效桥梁，不仅因为其在绿色发展指数评分方面取得了实质性进展，而且还因为其在绿色技术方面的规模和比较优势。在过去 10 年中，中国已成为可再生能源发展的全球领导者，其在提高能源效率方面的成功故事甚至可以追溯至更早的时候。虽然沙特阿拉伯和塞浦路斯等国家也有很大改善，但中国的规模使其在向其他不同规模的“一带一路”国家提供成功经验方面具有独特性。此外，中国不仅在绿色技术发展方面表现良好，而且实现了前所未有的经济发展，这使其对欠发达的“一带一路”国家更具吸引力。由于地理位置接近，历史关系和类似起点，中国可以为“一带一路”国家提供世界一流的绿色技术、技术知识和成功的政策经验。绿色发展指数评分系统可以使这项技术和政策传播过程更加透明和有影响力。

第五章　结论与建议

“一带一路”绿色发展极为重要。为了促进“一带一路”的绿色发展，本报告制定了绿色发展指数，以平衡自然资源、绿色技术和发展成果三个绿色发展的维度。绿色发展指数可以用于分析现有的绿色发展水平、变化方向和变化原因，从而加强“一带一路”绿色发展的相关国际合作。

本报告的研究仍存在一些需要改进的问题。其中，数据是我们面临的最大限制因素。在研究过程中，我们尝试建立一个更加系统和全面的指标体系，例如在自然资源中包括更多的资源类别，包括更具体的绿色技术，并考虑是否有更好的指标来描述绿色发展的结果。但数据可用性限制了我们构建更好的指标系统的能力。全面、公开、透明的评价体系，包括指标和指数，是推动绿色“一带一路”的必要条件。未来，为了推动“一带一路”倡议，我们应该吸引更多的国际组织，如联合国、世界银行等，为“一带一路”的绿色发展制定更具体的指标，提高数据质量，建立开放性数据共享平台。

本报告是评估“一带一路”倡议绿色发展的有益尝试。未来中国人民大学绿色发展指数研究小组将继续更新和改进此项研究，以跟踪“一带一路”绿色发展的进展。通过实施与“一带一路”相关的更多合作项目，有望就“一带一路”如何在未来促进绿色发展提出更具体的建议。

参考文献

Angang Hu, 2017, *China: Innovative Green Development*, China Renmin University Press.

Andrew Scott, William McFarland and Prachi Seth, 2013, "Research and Evidence on Green Growth".

China Green Development Index Research Group, 2017, *China Green Development Index Report* 2016: *Regional Comparison*, Beijing Normal University Press.

Chowdhury, S., Squire, L., 2006, "Setting Weights for Aggregate Indices: An Application to the Commitment to Development Index and Human Development Index", *Journal Development Studies*, 42, 761-771.

Dual Citizen LLC, 2016, "The Global Green Economy Index (GGEI 2016)".

Global Green Growth Institute, 2012, "Green Growth in Practice-Lessons from Country Experiences".

OECD, 2011, “Toward Green Growth”.

OECD, 2017, “Green Growth Indicators 2017”.

UNEP, 2011, “Towards a Green Economy: Pathways to Sustainable Development and Poverty Eradication”.

World Bank, 2013, “Seizing the Opportunity of Green Development in China”.

邹巅、廖小平：《绿色发展概念认知的再认知——兼谈习近平的绿色发展思想》，《湖南社会科学》2017年第2期。

王玲玲、张艳国：《“绿色发展”内涵探微》，《社会主义研究》2012年第5期。

张友国：《公平、效率与绿色发展》，《求索》2018年第1期。

唐啸、胡鞍钢：《绿色发展与“十三五”规划》，《学习与探索》2015年第11期。

侯伟丽：《21世纪中国绿色发展问题研究》，《南都学坛》2004年第3期。

许勤华：《中国能源生产与消费取向：自发达国家行为观察》，《改革》2014年第8期。

致　谢

感谢中国人民大学环境学院王克副教授对本研究的大力支持，从规划、调研到数据源获取，都提供了宝贵的建议和经验分享；感谢中国人民大学公共管理学院董长贵博士，在本研究方法论的构建上，给予了最前沿的思考和技术支持；感谢中国人民大学环境学院龚亚珍副教授的积极参与；感谢生态环保部气候中心王际杰，中国人民大学环境学院本科生项启昕、张婉琳、夏侯沁蕊、朋晶和国际关系学院博士生袁淼的帮助，协助进行了文献及数据整理工作。特别感谢美国哥伦比亚大学全球能源政策中心大卫·桑德罗先生及其团队对本研究提出的非常有价值的建议。

许勤华

2020年1月16日

Contents

Introduction

China has laid out an overall opening-up plan through the development of Silk Road Economic Belt and the 21st-Century Maritime Silk Road (hereafter referring as the Belt and Road Initiative or BRI). However, some major fossil fuel producers and consumers as well as many ecologically fragile countries are both involved in BRI. Thus, it is highly important to minimize the negative ecological impacts brought by the development of the Belt and Road and avoid a traditional development pathway characterized with "polluting first and cleaning up later". As such, integrating the idea of green development into the development of the Belt and Road should become a key component in the top design. BRI Green Development has great significance both for the successful implementation of BRI and for world's green and low carbon transition as well as successful implementation of global SDGs. BRI green development can help other developing countries avoid reliance on traditional high-carbon growth models and pursue more efficient and innovative paths which is demonstrated by China.

Chinese government has always advocated BRI green development. At the Belt and Road Summit held in May 2017, China's President Xi Jinping advocated that countries should jointly pursue new idea of green development, promote a lifestyle that is green, low-carbon, circular and sustainable, and strengthen the cooperation among countries in ecological conservation and environmental protection in order to achieve a common goal of sustainable development by 2030. Four ministries of China's central government also jointly released "Guiding Opinions on Promoting the Construction of Green Belt and Road" in May 2017.

According to the World Bank, "green development is a pattern of development that decouples growth from heavy dependence on resource use, carbon emissions and environmental damage, and promotes growth through the creation of new green product markets, technologies, investments, and changes in consumption and conservation behavior". But a comprehensive green development concept needs to include both development processes and outcomes. So we integrate the concept of “unity of nature and humanity” in traditional Chinese philosophy, the Marxist dialectics of nature and the contemporary theory of sustainable development to development the concept of green development in the context of BRI as figure 1 shows. Green Development will include a process of development which being closely linked with “green growth” and “sustainable development” and a stage of development which being closely linked with “green economy”.

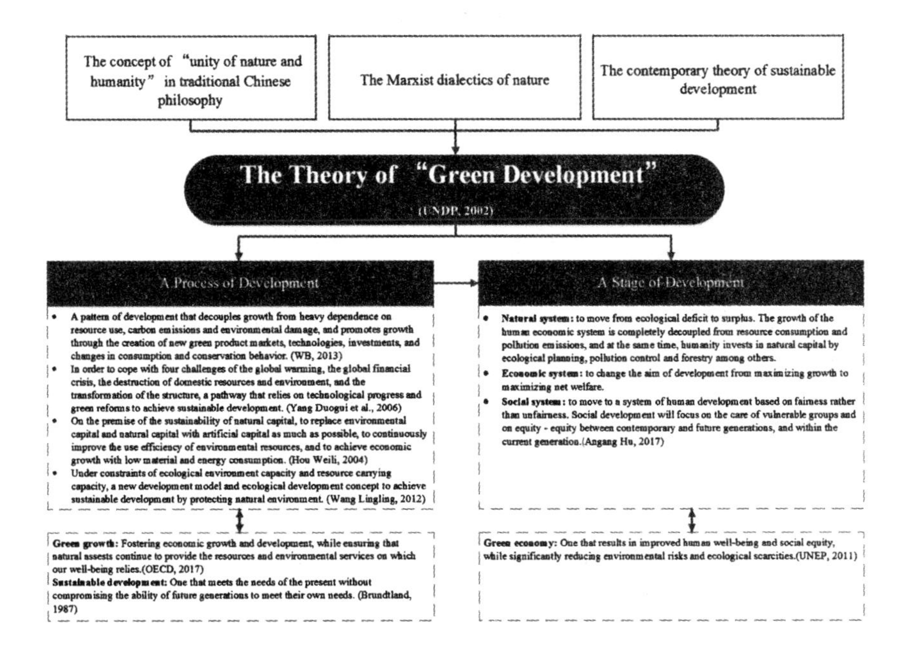

Figure 1 Concept of Green Development in the Context of BRI

Source: UNDR, 2002.

Following the above concept of Green Development in the context of BRI, Renmin University of China has established amultidisciplinary research team covering international relations, public administration, energy governance, environmental economics and climate change issues to conduct a research on BRI Green Development Index (GDI) . Research team has developed an index system on BRI Green Development toa ssess thel evel of green development in BRI countries and identify gaps and the main influencing factors causing gaps. This assessment can be used to support the identification of key areas and technologies that contribute to the level of green development of BRI countries.

The rest of this report introduces the design of GDI, data sources, weighting methods, research results, policy implications, conclusions and suggestions for further work.

1. Green Development Index System

1.1 Framework

By definition, "development" can mean both a process of making progresses and a result of achieving a certain level of economic, societal and environmental outcome. So is the green development. In this index system, the green development is divided into two development stages and a development process (see Figure 1) . The natural assets, which pose constrains or lay the foundation to the future development, is the endowment and initial stage of an economy. The development outcomes, which may accumulate by time, are the results and current state of the overall green development. In the process of turning natural assets into development outcomes, the green technologies act as an important role in shaping the economy and leading the development direction. These three categories are integrated components in the whole process of green development, thus all included in the Green Development Index System.

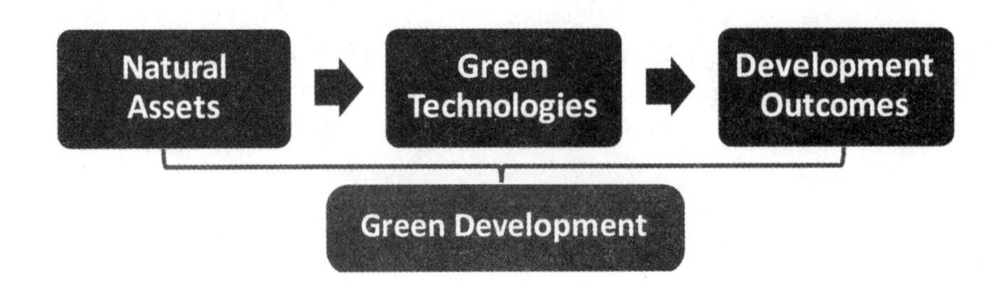

Figure 2 Green Development Index Framework

1.2 Categories and Indicators

Under the framework, a database is established in support of the index system. It covers 98 countries, including 65 countries under the BRI, China, all OECD countries, as well as BRI-relevant countries. The index system is consisted of 3 categories, 15 sub-categories and 20 indicators, with a time span from 2006 to 2015. Total scale indicators and average level indicators are both included, considering the balance between internal development of small countries and spillover effect of world powers. Data are mostly retrieved from official and reliable sources, including World Bank, UNDP, UNEP, IEA, IRENA, UNESCO, Yale University, and UNU etc.. Detailed information is shown in table 1.

Table 1 **The Green Development Index System for the BeltandRoad Countries**

Category	Sub-Category	Indicator and Unit	Indicator Influence	Indicator Type	Data Source
Natural Assets	Forest resource	Forest area (% of land area)	Positive	Average Level	WB
	Biodiversity	Biodiversity and habitat (Standardized, EPI score)	Positive	Average Level	EPI
	Water resource	Renewable internal freshwater resources, average (m^3/capita)	Positive	Average Level	WB
	Natural resources	Total natural resources rents (USD)	Positive	Total Scale	WB
		Total natural resources rents (% of GDP)	Positive	Average Level	
Green Techn-ologies	Renewable energy generation	Renewable electricity output (GWh)	Positive	Total Scale	IEA
		Renewable electricity output (% of total output)	Positive	Average Level	
	Renewable energy capacity	Renewable electricity capacity (GW)	Positive	Total Scale	UN
		Renewable electricity capacity (% of total)	Positive	Average Level	
	Energy efficiency	Energy consumption per GDP per capita (kg toe /USD)	Negative	Average Level	WB
	Green transportation	CO_2 emission in transportation, average (kg/capita)	Negative	Average Level	IEA
	Green building	CO_2 emission in building, average (kg/capita)	Negative	Average Level	IEA
	R&D technology Competitiveness	Research and development expenditure (million USD)	Positive	Total Scale	WB
		Research and development expenditure (% of GDP)	Positive	Average Level	

续表

Category	Sub-Category	Indicator and Unit	Indicator Influence	Indicator Type	Data Source
Development Outcomes	HDI	Human Development Index	Positive	Average Level	UNDP
	Inequality	Gini index	Negative	Average Level	UNU
	Access to Electricity	Access to electricity (% of population)	Positive	Average Level	WB
	CO_2	CO_2 emissions from fuel combustion, average (kg/capita)	Negative	Average Level	IEA
	PM2.5	Annual mean concentration (mg/m^3)	Negative	Average Level	EPI
		Average exposure to PM2.5 (Standardized, EPI score)	Positive	Average Level	

Notes: WB: World Bank; EPI: Environment Performance Index (Yale Center for Environmental Law & Policy, Yale University Center for International Earth Science Information Network, Columbia University in collaboration with the World Economic Forum); IEA: International Energy Administration; UN: United Nations; OECD: Organization for Economic Cooperation and Development; UNDP: United Nations Development Programme; UNU: United Nations University.

2. Data and Weighting

2.1 Data Standardization

The Green Development Index (GDI) for the Belt and Road countries is a comprehensive index to evaluate the natural assets, green technology and development achievements of various BRI countries. It consists of three sub-indices, each of which is synthesized by several quantitative indicators. Due to the existence of both positive indicators (the greater the value, the higher the level of green development) and negative indicators (the smaller the value, the higher the level of green development), as well as the great difference of magnitude and unit among the indicators, before calculating the sub-indices, it is necessary to standardize these indicators in order to make the overall calculation possible. Only in this way can the overall index be comparable not only horizontally across countries, but also vertically over time. While comparing the relative level of green development in different countries, the historical development process of the same country's green development level can also be investigated. Therefore, the standardization of the raw data is essential.

Data standardization has taken different approaches according to data characteristics.

For index with lower order of magnitude, dimensionless processing is carried out directly. The formulas for this kind of indicators are as follows:

$$Z_i = \frac{X_i - min\ (X_i)}{max\ (X_i) - min\ (X_i)} \times 100\% \qquad (1)$$

$$Z_i = \frac{max\ (X_i) - X_i}{\max(X_i) - min\ (X_i)} \times 100\% \qquad (2)$$

X_i is the value of each year of i index, $min\ (X_i)$ is the minimum value of the base year of i index, and $max\ (X_i)$ is the maximum value of the base year of i index.

For indicators with a relatively high order, the logarithmic value of the original value will be taken before the process of dimensionless:

$$Z_i = \frac{Ln\ X_i - min\ (Ln\ X_i)}{max(Ln\ X_i)\ - min\ (Ln\ X_i)} \times 100\% \qquad (3)$$

$$Z_i = \frac{max(Ln\ X_i) - Ln\ X_i}{max(Ln\ X_i) - min\ (Ln\ X_i)} \times 100\% \qquad (4)$$

2. 2 Weighting

Weights determine the relative importance of each indicator in the index system, and thus the performance ranking of those countries being evaluated. The role of weights is to combine various indicators into one coherent green development index. Although we have collected extensive data and made sure that all indicators listed in our GDI system (see Table 1) are related to green development, the correlations between each of those indicators and the unobserved green development index can be very different. Different weights tend to emphasize different aspects of green development, and there may not be an optimal weighting system.

There exist multiple ways to assign weights to indicators in an index system: some are more subjective such as the Delphi method and the analytic hierarchy process, and some are more objective such as the principal component analysis and the factor analysis. Furthermore, when it becomes difficult to differentiate two indicators, assigning equal weights to indicators is also common, such as the Human Development Index.

In this book, we have leveraged the statistical technique of factor analysis to come up with weights for the 20 indicators listed in Table 1. In using factor analysis, we have limited the analysis to extract the first three and most important common factors out of the 20 indicators in our index system. Based on the correlations between those indicators and the three common factors, it is interesting to find out these three common factors can be named as development, technology and environment, respectively, which are very similar to the three key components of our green development definition. This gives us more confidence to our theoretical definition of green development in the first place, also showing the power of factor analysis.

The final weights for the 20 GDI indicators are shown in Table 2 below. According to the results, although development (outcomes) is the most important common factor being extracted out, (green) technology is the category with the highest weight of 56. 3. This is because several indicators in the green technology category also correlate highly with the development common factor, such as CO_2 emission in building as well as Research and development expenditure, thus these indicators will obtain some weights from the development common factor.

Within the green technologies category, renewable energy development, R&D expenditure and energy efficiency are the three most important ones with the highest weights. Within the development outcomes category, HDI, CO_2 emissions and access to electricity are relatively more important. Lastly, within the natural assets category, water resource seems to matter more compared to other assets, highlighting the scarcity of freshwater resources in the Belt and Road countries.

2.3 Calculation of Sub-index and Synthesis of GDI

With the provided weights (W_j) above, one can easily calculate the GDI index and any sub-indices for each country and year. For example, for the total GDI index, we have:

$$GDI_{it} = \sum_j W_j X_{ijt} \tag{5}$$

where i indicates for a country, t for a year, and j for an indicator. As such, GDI_{it} is a weighted average of 20 different indicators X_{ijt} .

Similarly, we can calculate GDI-sub indices for natural assets, green technologies and development outcomes.

Table 2 **Weights for Indicators in the Green Development Index**

Category (Weight)	Sub-Category (Weight)	Indicator and Unit	Final Weight
Natural Assets (17.3)	Forest resource (3.6)	Forest area (% of land area)	3.6
	Biodiversity (3.4)	Biodiversity and habitat (Standardized, EPI score)	3.4
	Water resource (5.5)	Renewable internal freshwater resources, average (m^3/capita)	5.5
	Natural resources (4.9)	Total natural resources rents (million USD)	2.6
		Total natural resources rents (% of GDP)	2.3
Green Technologies (56.3)	Renewable energy generation (12.3)	Renewable electricity output (GWh)	8.0
		Renewable electricity output (% of total output)	4.2
	Renewable energy capacity (11.8)	Renewable electricity capacity (KW)	7.4
		Renewable electricity capacity (% of total)	4.4
	Energy efficiency (6.5)	Energy consumption per GDP per capita (kg toe /USD)	6.5
	Green transportation (5.0)	CO_2 emission in transportation, average (kg/capita)	5.0
	Green building (5.0)	CO_2 emission in building, average (kg/capita)	5.0
	R&D technology Competitiveness (15.7)	Research and development expenditure (million USD)	8.6
		Research and development expenditure (% of GDP)	7.1
Development Outcomes (26.4)	HDI (9.2)	Human Development Index	9.2
	Inequality (1.0)	Gini index	1.0
	Access to Electricity (4.6)	Access to electricity (% of population)	4.6
	CO_2 (1.6)	CO_2 emissions from fuel combustion, average (kg/capita)	6.5
	PM2.5 (5.0)	Annual mean concentration (mg/m^3)	1.6
		Average exposure to PM2.5 (Standardized, EPI score)	3.4

Notes: Weights in a column sum up to 100.

3. Calculation Results and Analyses

3.1 Countries Ranking based on GDI

Based on the above research methods, this study calculates the green development index and the sub-index values of the BRI countries. The results of the top 20 countries in 2015 (GDIs and their sub-indices) are shown in Tables 3①.

Table 3 **Top 20 of BRI Countries (GDI 2015)**

Ranking	Country	Sub-GDI Natural Assets	Sub-GDI Green Technology	Sub-GDI Development Outcome	GDI
1	Russia	70.9	57.6	81.7	66.3
2	Albania	53.5	68.2	69.7	66.1

① The values of the GDI and sub-indices in the table are all standardized.

续表

Ranking	Country	Sub-GDI Natural Assets	Sub-GDI Green Technology	Sub-GDI Development Outcome	GDI
3	China	60. 6	71. 1	57. 6	65. 7
4	Slovenia	73. 6	52. 9	84. 5	64. 8
5	Greece	64. 1	54. 7	82. 4	63. 7
6	Czech	62. 5	54. 0	83. 1	63. 2
7	Croatia	66. 5	54. 6	77. 1	62. 6
8	Kyrgyzstan	51. 7	64. 3	65. 5	62. 4
9	Poland	63. 3	53. 8	80. 2	62. 4
10	Malaysia	75. 3	50. 5	78. 4	62. 2
11	Latvia	70. 6	49. 9	79. 7	61. 3
12	Romania	62. 2	54. 7	74. 8	61. 3
13	Turkey	45. 9	60. 3	73. 3	61. 3
14	Slovakia	65. 8	50. 0	79. 7	60. 5
15	Lithuania	65. 1	48. 8	79. 3	59. 7
16	Estonia	70. 7	43. 4	87. 0	59. 6
17	Qatar	38. 0	55. 0	82. 4	59. 3
18	Georgia	61. 8	51. 8	71. 7	58. 8
19	Bulgaria	61. 0	49. 4	76. 4	58. 6
20	Israel	43. 6	50. 1	82. 3	57. 4

As the tables have shown, Russia, Albania and China occupied the top 3 of the lists. It is noteworthy that Russia obtained the first place, mainly because of its relatively high scoresin all of natural assets, green technologies, and development outcomes. China ranked third place in 2015, which is mainly due to the continuous improvement of green technology scores.

For comparison, the top 20 of OECD countries are listed in Table 4. The green development level of the Belt and Road countries differs from each other in a wider range than the OECD countries. The hugest gap between the Belt and Road countries was 28. 0, while that of the latter was only 22. 1. In 2015, regarding GDI, the top ten countries of the Belt and Road were Russia, Albania, China, Slovenia, Greece, Czech republic, Croatia, Kyrgyzstan, Poland, Malaysia, and their green development indexes were close to the average level of the OECD countries.

Table 4 **Top 20 of OECD Countries (GDI 2015)**

Ranking	Country	Sub-GDI Natural Assets	Sub-GDI Green Technology	Sub-GDI Development Outcome	GDI
1	Norway	69. 4	75. 9	92. 0	79. 0
2	Sweden	75. 7	73. 7	87. 3	77. 6
3	Canada	71. 5	73. 2	91. 0	77. 6
4	Switzerland	63. 5	78. 1	84. 8	77. 3
5	Austria	69. 5	75. 8	84. 9	77. 1
6	United States	66. 8	71. 6	90. 3	75. 7
7	Japan	74. 0	70. 8	86. 6	75. 5
8	Germany	61. 1	74. 5	87. 2	75. 5
9	Finland	79. 7	66. 5	89. 8	74. 9
10	Denmark	57. 0	70. 2	88. 2	72. 7
11	Australia	63. 3	63. 8	93. 5	71. 5
12	France	65. 7	67. 5	83. 5	71. 4
13	Iceland	59. 3	66. 7	88. 7	71. 2
14	Italy	66. 0	67. 7	80. 8	70. 8

续表

Ranking	Country	Sub-GDI Natural Assets	Sub-GDI Green Technology	Sub-GDI Development Outcome	GDI
15	New Zealand	73.0	61.4	88.5	70.5
16	Spain	65.6	65.7	84.0	70.5
17	Korea	65.4	66.5	79.4	69.7
18	Portugal	64.1	63.8	81.4	68.5
19	Ireland	58.0	62.4	88.2	68.4
20	Belgium	58.4	63.8	83.8	68.1

As can be seen from the table, compared with BRI top 20, OECD countries have relatively higher GDIs. Actually, a comparative analysis of GDI time series between BRI countries and OECD countries shows that the annual average index (2006 – 2015) of OECD is 67.4, and that of BRI countries is only 54.3.

The Belt and Road Green Development Index (GDI) consists of sub-indexes of three dimensions: natural assets, green technologies and development outcomes. According to three sub-index scores, theresults of the top 10 countries were shown in Table 5.

In 2015, the top ten countries on the natural assets index were Malaysia, Slovenia, Russia, Estonia, Latvia, Indonesia, Cambodia, Brunei, Croatia and Slovakia. The top ten countries in the green technologies sub-index were China, Albania, Kyrgyzstan, Turkey, Russia, India, Turkmenistan, Qatar, Greece and Tajikistan. The top ten countries in the development outcomes sub-index were Singapore, Estonia, Brunei, Slovenia, Saudi Arabia, the Czech Republic, Cyprus, Kazakhstan, Greece and Qatar. Among the three sub-indexes, there was a widest gap between the BRI countries and the OECD countries especially in green technologies sub-index. Likewise, the growth of GDI in the OECD countries was mainly attributed to the rising of the green technologies sub-index. During the period of 2006 to 2015, among the top ten GDI increasing countries, it was the increase of green technologies index that contributed to the growth of the whole, reaching to an average of 72.0%. This indicates that the promotion and application of green technologies are not only the weak point of the green development of the participating countries, but also a key area which has the greatest potential for improvement. It is of great importance that we should cooperate more with the Belt and Road countries in technology and accelerate the promotion and application of green technologies and products, which can improve the green development of the partner countries.

Table 5 **Top 10 of BRI Countries (Sub-GDIs 2015)**

Ranking	Country	Sub-GDI Natural Assets	Country	Sub-GDI Green Technology	Country	Sub-GDI Development Outcome
1	Malaysia	75. 29	China	71. 12	Singapore	87. 06
2	Slovenia	73. 56	Albania	68. 24	Estonia	87. 05
3	Russia	70. 89	Kyrgyzstan	64. 32	Brunei	86. 93
4	Estonia	70. 69	Turkey	60. 34	Slovenia	84. 50
5	Latvia	70. 55	Russia	57. 58	United Arab Emirates	83. 47
6	Indonesia	68. 71	India	57. 01	Czech	83. 11
7	Cambodia	67. 00	Turkmenistan	56. 63	Cyprus	83. 02
8	Brunei	66. 71	Qatar	54. 98	Kazakhstan	82. 90
9	Croatia	66. 54	Greece	54. 71	Greece	82. 43
10	Slovakia	65. 76	Tajikistan	54. 67	Qatar	82. 38

3. 2 Comparison Analysis

As the figure 3 shows, BRI countries have a lower GDI averagea nd generally more scattered distribution pattern compared to OECD countries. In 2015, BRI countries have a GDI range of 28. 0, while OECD countries only have a gap of 22. 1.

Also, obvious variances of GDI scores among geographical groups exist (see figure 4) . In 2015, 15 countries in Central East Europe have the highest GDI average of 60, followed by 5 countries in Central Asia with GDI averaging 56.

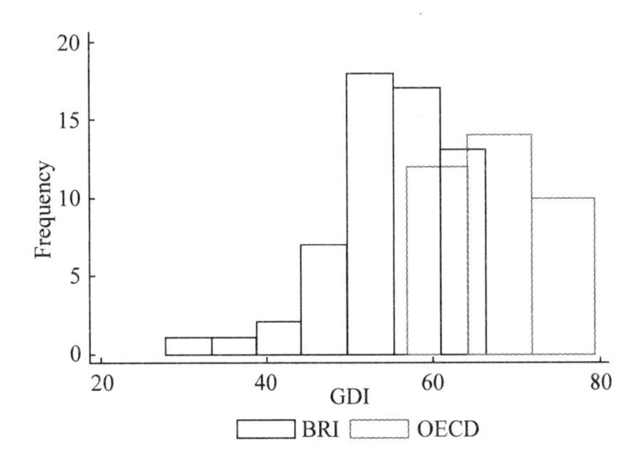

Figure 3 GDI Distributions of BRI and OECD countries (2015)

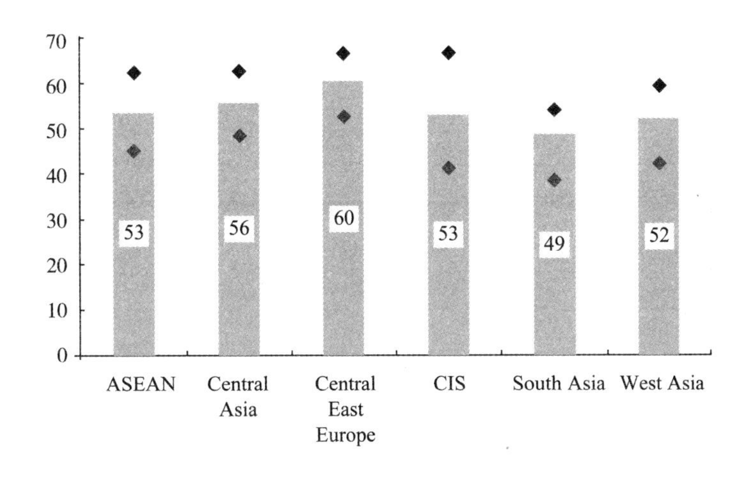

Figure 4 GDI 2015 by Geographical Groups

3.3 Trend Analysis of GDI

In the time span of our analysis, BRI countries have made huge progresses in green development that are well interpreted by our index system. The average GDI of BRI countries rises from 52.0 in 2006 to 54.9 in 2015 with a growth rate of 5.6%, faster than OECD countries (4.2%). As for the growing speed, the fastest ten countries in GDI were Saudi Arabia, Cyprus, China, Cambodia, Lithuania, Myanmar, Bulgaria, Estonia, Poland and Malaysia. Top 10 up-trending GDI of BRI countries have an average increase of 7.6, as China ranks the third. This increase was dominantly driven by the green technology development taking place in BRI countries, which contributes an average of 72.0% to the top 10 GDI improvements.

Table 6 **Top 10 BRI Countries with Fastest Growth of GDI**

Countries	GDI Improvement	Rank Improvement
Saudi Arabia	11.07	28
Cyprus	9.19	16
China	8.49	10
Cambodia	8.23	5
Lithuania	6.65	12
Myanmar	6.57	12
Bulgaria	6.49	10
Estonia	6.48	9
Poland	6.31	9
Malaysia	6.11	9

In addition, there is a strong correlation between GDI and GDP per capita, which can be approximately described as inverted U-shaped curve. Compared with OECD countries, the Belt and Road countries are mainly located on the left bottom of the curve, in which interval the GDI is more

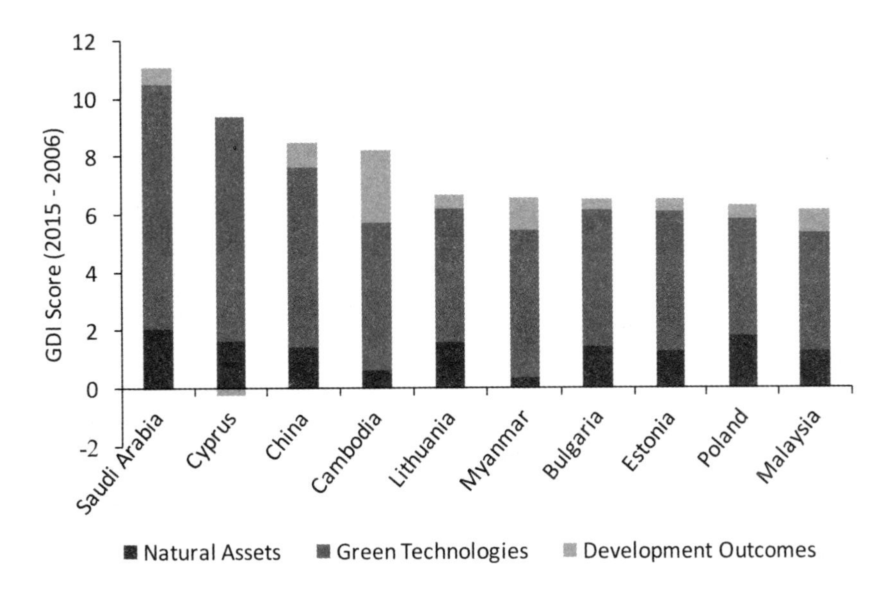

Figure 5 Comparison of the Contribution of Three Sub-GDIs for Growth of GDI, 2006 – 2015

correlated with GDP per capita. It shows that in the lower stage of GDP per capita, there is a consistent trend in direction between the economy development and green development. Therefore, based on the stage of development and characteristics of the BRI countries, the top priority of the partner countries is still to develop the economy. At the same time, the green development can be simultaneously upgraded with the economy development. However, in the process of economy development, countries along the Belt and Road need to abandon the traditional development path, namely "abate after pollution", and should transform to the green development way, relying on the advantage of backwardness of the green technologies.

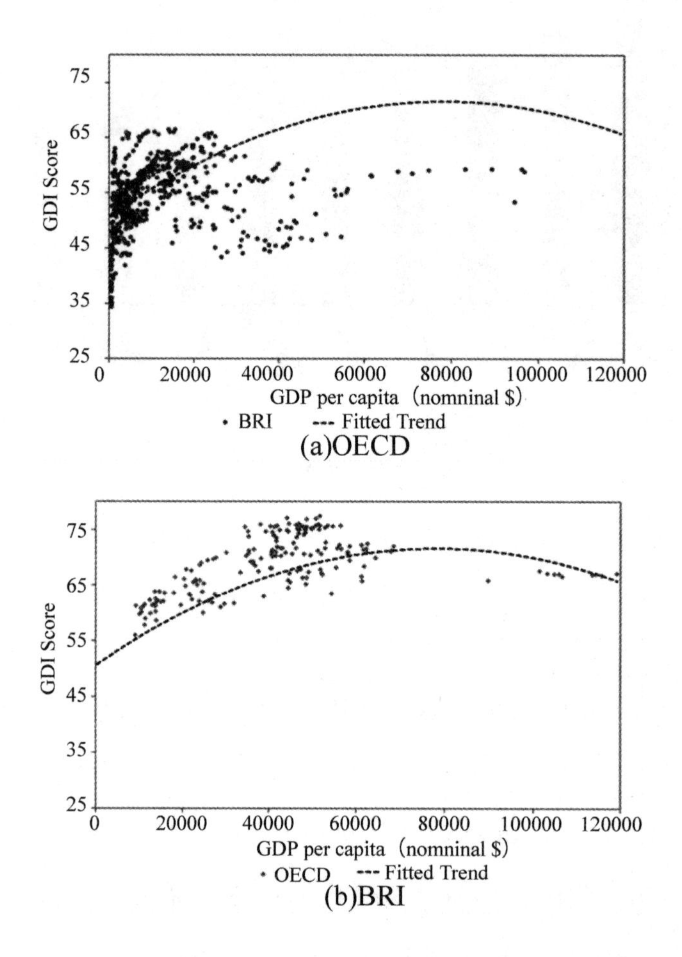

Figure 6 Kuznets Curve between GDI and GDP per capita

3.4 Trend Analysis of China's GDI

China witnessed a growth in GDI from 57. 2 in 2006 to 65. 7 in 2015 which is one of the fastest growing countries. The GDI level of China in 2015 ranked No. 3 among the countries along the Belt and Road, which was close to the average level of OECD countries in 2006. Besides, the green technologies sub-index of China is 71. 1, ranking first among the Belt and Road countries. The increase in the green technologies sub-index mainly explained for the significant increase in GDI of China, which is due to China's great efforts in research, demonstration and promotion of green low-carbon technologies, such as energy efficiency and renewable energy. The distinct rise in GDI levels of China in recent years reflects that China, as the largest developing country, is exploring and opening up a green sustainable development way through innovative development paths gradually, which is suitable for developing countries. It provides models and experiences that can be used for reference for other developing countries. In some areas, such as green infrastructurec onstruction and standard setting, green technologies research, demonstration and promotion. There is an immense potential for cooperation between China and the Belt and Road countries.

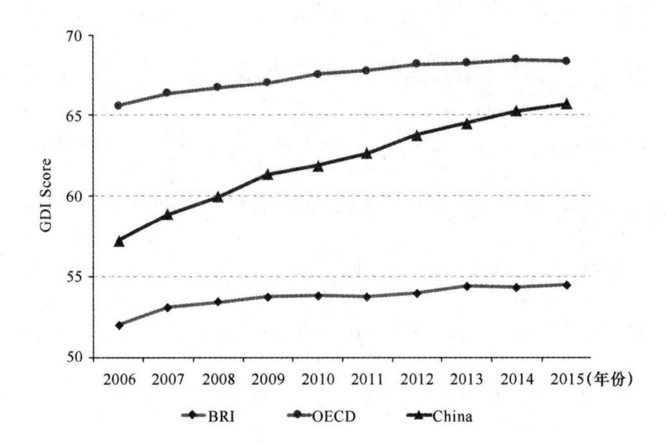

Figure 7 GDI Score Increases Over Time of China Comparing with BRI and OECD Countries

4. Policy Implications of GDI Analysis

The above analyses of GDI results show that the performance of green development varies from one country to another; this is especially true to BRI. For BRI countries in 2015, while the highest GDI score is 66, the lowest score is only 28, a difference of almost 40. For OECD countries, the difference is only 22, with the highest score as 79 and the lowest as 57. Therefore, BRI countries not only face the task of economic development, but also are less green in general when compared to OECD countries. Although some BRI countries such as Russia and Albania have arrived a development stage close to certain OECD countries, their development status has not changed much during the last ten years.

The low starting points of BRI countries also provide

great green development potentials for them. When looking at the GDI score changes over time, several countries stand out including Saudi Arabia, Cyprus and China, all starting from low GDI scores. Interestingly, their improvements point to the same underlying reason—green technologies, though the specific technologies tend to be slightly different: Saudi Arabia improves due to its R&D development and energy efficiency, whereas Cyprus and China improve due to its renewable energy development and energy efficiency. This suggests multiple channels for a country to climb up the GDI ladder, learning from the successful experience in other countries. As a result, adding green technology into the concept of green development is not only necessary, but also useful to suggest pathways for a country to achieve better green development.

Another important policy implication of our GDI score analyses is that green development is not equivalent to development outcomes, with the latter including things like GDP per capita, or more broadly the Human Development Index. In general, there is an inverted U-shaped relationship between GDI and GDP per capita, meaning when GDP per capita arrives at a certain level, GDI score stops increasing. This holds true for both OECD countries and BRI countries. Within BRI, countries such as Singapore, Brunei and United Arab Emirates perform well in terms of development outcomes; however, all of them score poorly in the overall GDI. In order to tackle the global climate change challenge, BRI countries should adopt a more comprehensive concept of development and start to embrace the idea of green development.

China can be a useful bridge between BRI countries and OECD countries, not only because of its substantial improvements in GDI scores, but also because of its scale and comparative advantages in green technologies. Within the past decade, China has become a global leader in renewable energy development and its successful story in improving energy efficiency even dates back earlier. Although other countries such as Saudi Arabia and Cyprus have also improved a lot, the scale of China makes it unique in providing successful experiences to other BRI countries with varying sizes. Furthermore, China not only has performed well in green technologies development, but also has achieved unprecedented economic development, making it more attractive to under-developed BRI countries. Due to the geographical closeness, historical ties, and similar starting points, China can provide world-class green technologies, technology know-how and successful policy experiences to

BRI countries. Our GDI scoring system could help make this technology and policy diffusion process more transparent and impactful.

5. Conclusions and Suggestions

There is the importance of the green development of the Belt and Road. In order to promote the green development of BRI, we developed this Green Development Index in a good balance of natural assets, green technologies and development outcomes, the three dimensions of green development. GDI can be used to analyze the existing green development level, the direction of change and the main causesof the change, so as to strengthen the relevant cooperation of the green development of BRI.

There are still someissues in our research that need to be improved. Among them, data is the biggest constraint we face. During the research, we have tried to establish a more systematic and comprehensive indicator system, such as including more resource categories in natural resource assets, including more specific technologies in green technologies, and considering whether there are better indicators to describe the outcome of green development. But data availability limits our ability to build a better indicator system. A comprehensive, open and transparent evaluation system including indicators and indexes is necessary for the promotion of green Belt and Road. In the future, for the promotion of the Belt and Road Initiative, we should attract more international organizations, such as UN, WB, etc. , to set up more specific indicators for green development of BRI, improve data quality, and establish an open data sharing platform.

This book is a useful attempt to evaluate the green development of the BRI. The GDI Research Group of Renmin University will continue to update and improve this research in the future to track the progress of the green development of the Belt and Road. With the implementation of more collaborative projects related to BRI, we can provide more specific advice on how BRI can promote green development in the future.

References

Angang Hu, 2017, *China: Innovative Green Development*, China Renmin University Press.

Andrew Scott, William McFarland and Prachi Seth, 2013, "Research and Evidence on Green Growth".

China Green Development Index Research Group, 2017, *China Green Development Index Report* 2016: *Regional Comparison*, Beijing Normal University Press.

Chowdhury, S., Squire, L., 2006, "Setting Weights for Aggregate Indices: An Application to the Commitment to Development Index and Human Development Index", *Journal Development Studies*, 42, 761 – 771.

Dual Citizen LLC, 2016, "The Global Green Economy Index (GGEI 2016)".

Global Green Growth Institute, 2012, "Green Growth in Practice—Lessons from Country Experiences".

OECD, 2011, "Toward Green Growth".

OECD, 2017, " Green Growth Indicators 2017".

UNEP, 2011, "Towards a Green Economy: Pathways to Sustainable Development and Poverty Eradication".

World Bank, 2013, "Seizing the Opportunity of Green Development in China".

Appendix
Green Development Database*

* Note: GDP per capital (USD); Population (million); Forest area (% of land area); Biodiversity & Habifat (pre-standardized); Water resources [(per capita) / (m^3/capital)]; National resources (Total) (million USD); Natural resources (% of GDP); Renewable power generation (Total) (GWH); Renewable power generation (% of total output); Renewable capacity (Total) (KW); Renewable capacity (% of total capacity); Energy efficiency (kgtoe/USD); Green transportation (percapita) (kg CO_2 per capital); Green Building (per capita) (kg CO_2 per capital); R&D Technology Competifiveness (% of GDP); Access to electricity (% of population); CO_2 (per capita) (kg CO_2); PM2.5 concentration (mg/m^3) Air pollution-average exposure to PM2.5 (pre-standardized).

1. Austria

Year	id	along BR	OECD	income group	GDP per capita	Population	Forest area	Biodiver-ity & Habitat	Water resources	Natural resources	Natural resources	Renewable power generation	Renewable power generation
Year	id	brc	oecd	inc_ group	gdppc	pop	forest	bio	water	natural	natural_ pct	renew_ g	renew_ g_ pct
2005	241	0	1	high	38403	8. 23	46. 63			518. 73	0. 16	42521. 78	63. 63
2006	242	0	1	high	40635	8. 27	46. 66		6666. 22	679. 60	0. 20	42379. 37	65. 52
2007	243	0	1	high	46856	8. 30	46. 68	94. 35	6630. 11	895. 14	0. 23	44763. 34	68. 69
2008	244	0	1	high	51709	8. 32	46. 70	94. 32	6608. 95	1118. 03	0. 26	46291. 93	69. 23
2009	245	0	1	high	47963	8. 34	46. 72	94. 45	6587. 80	680. 08	0. 17	49149. 70	71. 14
2010	246	0	1	high	46858	8. 36	46. 75	94. 33	6566. 64	843. 43	0. 22	47095. 45	66. 21
2011	247	0	1	high	51375	8. 39	46. 78	94. 08	6545. 48	1103. 30	0. 26	43192. 70	65. 63
2012	248	0	1	high	48568	8. 43	46. 80	93. 74	6524. 32	1048. 81	0. 26	54108. 93	74. 53
2013	249	0	1	high	50717	8. 48	46. 84	93. 53	6479. 91	1015. 57	0. 24	53300. 29	78. 02
2014	250	0	1	high	51705	8. 55	46. 86	93. 5	6435. 49	892. 46	0. 20	53045. 24	81. 06
2015	251	0	1	high	44207	8. 64	46. 88	93. 54	6428. 46	516. 78	0. 14	49947. 71	76. 49
2016	252	0	1	high	44731	8. 74				494. 47	0. 13		

Year	Renewable capacity	Renewable capacity	Energy efficiency	Green transportation	Green Building	R&D Technology Competitiveness	R&D Technology Competitiveness	HDI	Inequality	Access to electricity	CO_2	PM2. 5 concentration	Air pollution-average exposure to PM2. 5	Missing data 15-in total	Percentage of overall missing data
	renew_ c	renew_ c_ pct	ee	trans	bldg	rd	rd_ pct	hdi	inequality	electricity	co2	pm_ con	pm_ expo		
2005	11902. 00	62. 98	0. 11			7530. 68	2. 38	0. 86	28. 72	100. 00		15. 31		0	9. 6%
2006	12080. 00	63. 16	0. 10	23. 30	1404. 40	7981. 34	2. 38	0. 88	29. 59	100. 00	8675. 50	15. 28			
2007	12277. 00	64. 28	0. 09	23. 60	1176. 90	9466. 19	2. 44	0. 88	30. 58	100. 00	8356. 30	15. 24	68. 62		
2008	12758. 00	62. 09	0. 08	22. 30	1257. 60	11122. 89	2. 58	0. 89	30. 45	100. 00	8339. 20	15. 21	70. 79		
2009	12963. 00	62. 35	0. 08	21. 60	1147. 70	10465. 71	2. 62	0. 90	31. 50	100. 00	7626. 10	15. 18	70. 78		
2010	13294. 00	62. 75	0. 09	22. 30	1189. 20	10715. 88	2. 73	0. 90	30. 25	100. 00	8214. 40	15. 15	76. 27		
2011	13843. 00	61. 37	0. 08	21. 80	1003. 00	11547. 26	2. 68	0. 90	30. 80	100. 00	7961. 40	14. 93	76. 67		
2012	14221. 00	62. 05	0. 08	21. 70	971. 20	11995. 83	2. 93	0. 90	30. 48	100. 00	7584. 30	15. 25	79. 34		
2013	14886. 00	63. 10	0. 08	22. 80	963. 70	12744. 01	2. 96	0. 90	27. 00	100. 00	7574. 60	15. 94	73. 04		
2014	15656. 00	65. 10	0. 07	22. 20	858. 00	13524. 08	3. 06	0. 90	27. 60	100. 00	7097. 50	16. 47	75. 93		
2015	16241. 00	66. 44	0. 07	22. 70	878. 90	11736. 34	3. 07	0. 90	27. 20	100. 00	7200. 70	15. 03	74. 94		
2016								0. 91		100. 00		14. 97			

2. Brunei

Year	id	along BR	OECD	income group	GDP per capita	Population	Forest area	Biodiver-ity & Habitat	Water resources	Natural resources	Natural resources	Renewable power generation	Renewable power generation
Year	id	brc	oecd	inc_ group	gdppc	pop	forest	bio	water	natural	natural_ pct	renew_ g	renew_ g_ pct
2005	61	1	0	high	26338	0. 37	73. 81						0. 00
2006	62	1	0	high	31158	0. 37	73. 47		22974. 19				0. 00
2007	63	1	0	high	32708	0. 37	73. 13		22674. 89				0. 00
2008	64	1	0	high	37798	0. 38	72. 79	18. 33	22392. 59				0. 00
2009	65	1	0	high	27727	0. 38	72. 45	17. 72	22110. 29				0. 00
2010	66	1	0	high	31453	0. 39	72. 11	19. 59	21827. 99				0. 00
2011	67	1	0	high	41787	0. 39	72. 11	34. 31	21545. 69			2. 00	0. 05
2012	68	1	0	high	41808	0. 40	72. 11	33. 65	21263. 40			2. 00	0. 05
2013	69	1	0	high	39151	0. 41	72. 11	33. 44	20954. 65			2. 00	0. 05
2014	70	1	0	high	40980	0. 41	72. 11	33. 86	20645. 90			2. 00	0. 04
2015	71	1	0	high	36608	0. 42	72. 11	34. 14	20379. 66			2. 00	0. 05
2016	72	1	0	high income	26939	0. 42	72. 11	34. 4					

Year	Renewable capacity	Renewable capacity	Energy efficiency	Green transportation	Green Building	R&D Technology Competitiveness	R&D Technology Competitiveness	HDI	Inequality	Access to electricity	CO_2	PM2. 5 concentration	Air pollution-average exposure to PM2. 5	Missing data 15-in total	Percentage of overall missing data
	renew_ c	renew_ c_ pct	ee	trans	bldg	rd	rd_ pct	hdi	inequality	electricity	co2	pm_ con	pm_ expo		
2005								0. 84		100. 00		7. 36		4	49. 6%
2006			2802. 30	205. 40			0. 84		100. 001	9604. 008. 7	1				
2007				3011. 30	290. 30			0. 84		100. 00	18294. 40	8. 67	95. 27		
2008				2970. 90	299. 00			0. 84		100. 00	18953. 80	8. 41	95. 27		
2009				2898. 40	280. 90			0. 84		100. 00	19203. 30	8. 57	95. 27		
2010				2990. 80	253. 60			0. 85		100. 00	17455. 50	7. 87	94. 31		
2011				3216. 50	257. 60			0. 85		100. 00	17572. 50	9. 24	94. 31		
2012				3306. 40	244. 70			0. 85		100. 00	17161. 90	9. 20	94. 31		
2013				3225. 90	229. 60			0. 85		100. 00	16671. 30	9. 22	94. 31		
2014				3239. 30	213. 10			0. 85		100. 00	16055. 90	9. 37	94. 34		
2015				3207. 50	200. 70			0. 85		100. 00	14127. 80	9. 53	88. 73		
2016								0. 85		100. 00					

3. Cambodia

Year	id	along BR	OECD	income group	GDP per capita	Population	Forest area	Biodiver-ity & Habitat	Water resources	Natural resources	Natural resources	Renewable power generation	Renewable power generation
Year	id	brc	oecd	inc_ group	gdppc	pop	forest	bio	water	natural	natural_ pct	renew_ g	renew_ g_ pct
2005	49	1	0	lower middle	472	13. 27	60. 79			107. 81	1. 72	59. 00	6. 12
2006	50	1	0	lower middle	538	13. 47	60. 07		8949. 66	132. 03	1. 82	68. 00	5. 77
2007	51	1	0	lower middle	629	13. 68	59. 35	87. 52	8817. 92	193. 26	2. 25	68. 00	4. 56
2008	52	1	0	lower middle	743	13. 88	58. 63	87. 69	8686. 62	319. 65	3. 10	65. 00	4. 39
2009	53	1	0	lower middle	735	14. 09	57. 91	87. 54	8555. 31	296. 65	2. 86	68. 00	5. 37
2010	54	1	0	lower middle	783	14. 31	57. 18	87. 59	8424. 01	301. 93	2. 70	55. 00	5. 50
2011	55	1	0	lower middle	879	14. 54	56. 46	87. 41	8292. 71	389. 91	3. 05	75. 00	7. 08
2012	56	1	0	lower middle	947	14. 78	55. 74	87. 06	8161. 41	321. 21	2. 30	540. 00	37. 66
2013	57	1	0	lower middle	1025	15. 02	55. 02	86. 85	8029. 42	288. 57	1. 87	1030. 00	57. 93
2014	58	1	0	lower middle	1095	15. 27	54. 30	86. 61	7897. 43	422. 46	2. 53	1872. 00	61. 14
2015	59	1	0	lower middle	1159	15. 52	53. 57	86. 59	7766. 54	313. 89	1. 75	2041. 00	46. 42
2016	60	1	0	lower middle	1270	15. 76	53. 57			380. 82	1. 90		

Year	Renewable capacity	Renewable capacity	Energy efficiency	Green transportation	Green Building	R&D Technology Competitiveness	R&D Technology Competitiveness	HDI	Inequality	Access to electricity	CO_2	PM2. 5 concentration	Air pollution-average exposure to PM2. 5	Missing data 15-in total	Percentage of overall missing data
	renew_ c	renew_ c_ pct	ee	trans	bldg	rd	rd_ pct	hdi	inequality	electricity	co2	pm_ con	pm_ expo		
2005	12. 96	5. 60	0. 55					0. 50		20. 50		8. 84	92. 27	0	17. 9%
2006	13. 96	4. 64	0. 47	117. 40	20. 30			0. 51		28. 08	217. 60	9. 07	90. 80		
2007	13. 96	4. 43	0. 40	143. 20	18. 00			0. 52	41. 14	30. 18	260. 00	8. 79	88. 50		
2008	14. 39	3. 73	0. 34	146. 10	9. 60			0. 52	35. 10	26. 40	261. 90	8. 68	88. 50		
2009	14. 35	3. 85	0. 49	188. 10	21. 00			0. 54	34. 02	34. 44	308. 80	8. 74	88. 50		
2010	15. 30	4. 23	0. 47	214. 90	20. 70			0. 55	32. 93	31. 10	321. 70	8. 77	82. 54		
2011	209. 06	36. 61	0. 43	223. 90	17. 90			0. 55	31. 85	38. 78	331. 90	9. 16	82. 54		
2012	227. 40	38. 95	0. 41	232. 40	19. 10			0. 56	30. 76	40. 96	340. 60	9. 59	82. 54		
2013	685. 40	59. 24	0. 39	233. 00	21. 30			0. 57	28. 02	43. 16	342. 80	9. 26	82. 54		
2014	931. 40	61. 55	0. 38	250. 50	22. 80			0. 57	27. 27	56. 10	397. 70	9. 33	77. 56		
2015	931. 70	56. 15	0. 37	271. 20	32. 40	21. 27	0. 12	0. 55	25. 85	47. 57	514. 10	9. 40	77. 56		
2016								0. 58		49. 77					

4. Canada

Year	id	along BR	OECD	income group	GDP per capita	Population	Forest area	Biodiver-ity & Habitat	Water resources	Natural resources	Natural resources	Renewable power generation	Renewable power generation
Year	id	brc	oecd	inc_ group	gdppc	pop	forest	bio	water	natural	natural_ pct	renew_ g	renew_ g_ pct
2005	253	0	1	high	36190	32. 31	38. 22			58647. 92	5. 02	372563. 63	60. 03
2006	254	0	1	high	40387	32. 57	38. 22		87571. 91	63377. 88	4. 82	363934. 09	59. 54
2007	255	0	1	high	44545	32. 89	38. 21	72. 29	86657. 94	66645. 90	4. 55	379095. 93	60. 30
2008	256	0	1	high	46596	33. 25	38. 20	72. 15	85728. 97	90092. 49	5. 82	388492. 23	61. 47
2009	257	0	1	high	40773	33. 63	38. 20	71. 95	84800. 00	33196. 30	2. 42	383233. 76	62. 85
2010	258	0	1	high	47447	34. 01	38. 19	72. 25	83871. 04	45145. 81	2. 80	370816. 15	61. 40
2011	259	0	1	high	52082	34. 34	38. 19	71. 89	82942. 07	58671. 18	3. 28	396741. 20	62. 34
2012	260	0	1	high	52497	34. 75	38. 18	72. 91	82013. 10	44530. 90	2. 44	402628. 19	63. 23
2013	261	0	1	high	52418	35. 15	38. 18	72. 82	81107. 47	46917. 29	2. 55	421157. 26	63. 30
2014	262	0	1	high	50633	35. 54	38. 17	72. 83	80201. 83	44705. 18	2. 48	419750. 74	62. 82
2015	263	0	1	high	43525	35. 83	38. 17	73. 14	79258. 70	14632. 66	0. 94	422712. 94	63. 01
2016	264	0	1	high	42349	36. 26				15543. 94	1. 01		

Year	Renewable capacity	Renewable capacity	Energy efficiency	Green transportation	Green Building	R&D Technology Competitiveness	R&D Technology Competitiveness	HDI	Inequality	Access to electricity	CO_2	PM2.5 concentration	Air pollution-average exposure to PM2.5	Missing data 15-in total	Percentage of overall missing data
	renew_ c	renew_ c_ pct	ee	trans	bldg	rd	rd_ pct	hdi	inequality	electricity	co2	pm_ con	pm_ expo		
2005	68083. 00	55. 26	0. 23			23076. 70	1. 97	0. 90		100. 00		8. 06		0	12. 1%
2006	69579. 00	55. 73	0. 20	158. 60	2466. 10	25671. 77	1. 95	0. 90		100. 00	16366. 90				
2007	70282. 00	55. 75	0. 18	166. 50	2540. 00	28023. 11	1. 91	0. 90	33. 90	100. 00	17150. 60		92. 59		
2008	71732. 00	56. 38	0. 18	166. 30	2446. 80	28871. 16	1. 86	0. 90	33. 83	100. 00	16361. 40		92. 59		
2009	73021. 00	55. 45	0. 19	162. 80	2326. 20	26312. 97	1. 92	0. 90	33. 75	100. 00	15302. 20		92. 59		
2010	74199. 00	56. 05	0. 16	168. 00	2177. 60	29698. 39	1. 84	0. 91	33. 68	100. 00	15540. 30	7. 27	92. 07		
2011	76268. 00	57. 36	0. 15	167. 80	2353. 00	32169. 37	1. 80	0. 91	33. 61	100. 00	15689. 90	7. 34	92. 07		
2012	77385. 00	59. 05	0. 15	169. 80	2160. 40	32784. 11	1. 80	0. 91	33. 53	100. 00	15514. 60	7. 30	92. 07		
2013	79192. 00	58. 96	0. 15	174. 40	2232. 50	31170. 63	1. 69	0. 92	33. 46	100. 00	15701. 40	7. 27	92. 07		
2014	81716. 00	59. 28	0. 16	174. 20	2263. 70	29066. 83	1. 62	0. 92	33. 39	100. 00	15598. 70	7. 21	91. 96		
2015	87759. 00	58. 76	0. 13	173. 80	2144. 80	25576. 73	1. 64	0. 91	33. 31	100. 00	15319. 50	7. 51	91. 96		
2016								0. 93		100. 00		7. 53			

5. China

Year	id	along BR	OECD	income group	GDP per capita	Population	Forest area	Biodiver-ity & Habitat	Water resources	Natural resources	Natural resources	Renewable power generation	Renewable power generation
Year	id	brc	oecd	inc_ group	gdppc	pop	forest	bio	water	natural	natural_ pct	renew_ g	renew_ g_ pct
2005	1	1	0	upper middle	1740	1303. 72	20. 56			115116. 54	5. 07	404451. 00	16. 18
2006	2	1	0	upper middle	2082	1311. 02	20. 72		2144. 85	150677. 69	5. 52	446882. 00	15. 59
2007	3	1	0	upper middle	2673	1317. 89	20. 88	78. 57	2134. 48	221447. 02	6. 29	500911. 00	15. 26
2008	4	1	0	upper middle	3441	1324. 66	21. 05	78. 21	2124. 11	447288. 12	9. 81	614990. 00	17. 74
2009	5	1	0	upper middle	3801	1331. 26	21. 21	77. 68	2113. 74	189034. 63	3. 74	663651. 00	17. 86
2010	6	1	0	upper middle	4515	1337. 71	21. 37	77. 49	2103. 37	377425. 61	6. 25	783647. 22	18. 62
2011	7	1	0	upper middle	5574	1344. 13	21. 53	77. 09	2093. 00	588984. 56	7. 86	790445. 03	16. 76
2012	8	1	0	upper middle	6265	1350. 70	21. 70	76. 93	2082. 63	419324. 73	4. 96	997123. 46	19. 97
2013	9	1	0	upper middle	6992	1357. 38	21. 86	76. 65	2072. 27	379433. 66	4. 00	1105585. 64	20. 30
2014	10	1	0	upper middle	7587	1364. 27	22. 03	76. 53	2061. 91	299798. 09	2. 90	1283961. 41	22. 61
2015	11	1	0	upper middle	7925	1371. 22	22. 19	76. 33	2051. 53	115512. 45	1. 06	1402101. 44	23. 93
2016	12	1	0	upper middle	8123	1378. 67	22. 19		126587. 66	1. 13			renew_ g_ pct

Year	Renewable capacity	Renewable capacity	Energy efficiency	Green transportation	Green Building	R&D Technology Competitiveness	R&D Technology Competitiveness	HDI	Inequality	Access to electricity	CO_2	PM2. 5 concentration	Air pollution-average exposure to PM2. 5	Missing data 15-in total	Percentage of overall missing data
	renew_ c	renew_ c_ pct	ee	trans	bldg	rd	rd_ pct	hdi	inequality	electricity	co2	pm_ con	pm_ expo		
2005	118450. 00	21. 17	0. 80			29722. 10	1. 31	0. 66	42. 50	97. 74		45. 70		0	9. 6%
2006	132360. 00	19. 83	0. 73	343. 30	293. 00	37433. 28	1. 37	0. 68	48. 70	98. 07	4518. 10	47. 30			
2007	152430. 00	19. 86	0. 61	366. 00	302. 10	48413. 06	1. 37	0. 69	48. 40	98. 40	4915. 70	49. 90	3. 27		
2008	180992. 00	20. 98	0. 49	392. 80	289. 20	65755. 42	1. 44	0. 69	49. 10	98. 75	4994. 70	49. 90	3. 27		
2009	213942. 00	22. 89	0. 47	397. 90	294. 20	84112. 91	1. 66	0. 71	49. 00	99. 12	5284. 60	49. 80	3. 27		
2010	245928. 00	23. 83	0. 43	434. 60	311. 30	103281. 46	1. 71	0. 71	52. 00	99. 70	5762. 20	48. 40	0. 00		
2011	281458. 00	24. 68	0. 37	472. 10	327. 90	133101. 87	1. 78	0. 72	47. 70	99. 75	6299. 00	49. 10	0. 00		
2012	314328. 00	25. 62	0. 34	516. 90	338. 50	161371. 33	1. 91	0. 73	48. 00	99. 92	6381. 90	48. 00	0. 00		
2013	372878. 00	27. 72	0. 32	554. 80	351. 10	188961. 04	1. 99	0. 74	50. 54	99. 98	6626. 20	49. 61	0. 00		
2014	426318. 00	28. 92	0. 29	572. 70	356. 90	209202. 46	2. 02	0. 74	51. 07	100. 00	6623. 60	49. 85	2. 31		
2015	492498. 00	29. 71	0. 18	610. 10	372. 20	224456. 40	2. 07	0. 71	51. 60	100. 00	6590. 10	50. 10	2. 31		
2016	renew_ c	renew_ c_ pct	ee	trans	bldg	rd	rd_ pct	hdi	inequality	electricity	co2	pm_ con	pm_ expo		

6. Estonia

Year	id	along BR	OECD	income group	GDP per capita	Population	Forest area	Biodiver-ity & Habitat	Water resources	Natural resources	Natural resources	Renewable power generation	Renewable power generation
Year	id	brc	oecd	inc_ group	gdppc	pop	forest	bio	water	natural	natural_ pct	renew_ g	renew_ g_ pct
2005	181	1	1	high	10338	1. 35	53. 13			112. 72	0. 80	111. 00	1. 09
2006	182	1	1	high	12595	1. 35	53. 04		9451. 91	131. 26	0. 77	130. 00	1. 34
2007	183	1	1	high	16586	1. 34	52. 96	100	9480. 26	137. 73	0. 62	145. 00	1. 19
2008	184	1	1	high	18095	1. 34	52. 87	100	9506. 04	181. 31	0. 75	197. 00	1. 86
2009	185	1	1	high	14726	1. 33	52. 79	100	9531. 82	149. 87	0. 76	541. 00	6. 16
2010	186	1	1	high	14641	1. 33	52. 70	100	9557. 60	209. 95	1. 08	1044. 00	8. 05
2011	187	1	1	high	17454	1. 33	52. 69	100	9583. 38	263. 84	1. 14	1179. 00	9. 14
2012	188	1	1	high	17491	1. 32	52. 68	100	9609. 16	295. 41	1. 28	1477. 00	12. 34
2013	189	1	1	high	19155	1. 32	52. 67	100	9638. 95	266. 56	1. 06	1220. 00	9. 19
2014	190	1	1	high	20148	1. 31	52. 66	100	9668. 74	290. 52	1. 10	1389. 00	11. 16
2015	191	1	1	high	17295	1. 32	52. 65	100	9692. 09	220. 24	0. 97	1502. 00	14. 42
2016	192	1	1	high	17737	1. 32	52. 65			232. 68	1. 00		

Year	Renewable capacity	Renewable capacity	Energy efficiency	Green transportation	Green Building	R&D Tec-hnology Competitiveness	R&D Technology Competitiveness	HDI	Inequality	Access to electricity	CO_2	PM2. 5 concentration	Air pollution-average exposure to PM2. 5	Missing data 15-in total	Percentage of overall missing data
	renew_ c	renew_ c_ pct	ee	trans	bldg	rd	rd_ pct	hdi	inequality	electricity	co2	pm_ con	pm_ expo		
2005	35. 00	1. 37	0. 37			129. 37	0. 92	0. 83	34. 10	100. 00		7. 35		0	9. 6%
2006	35. 00	1. 37	0. 30	1704. 20	328. 20	189. 43	1. 12	0. 84	33. 10	100. 00	11495. 20	7. 33			
2007	54. 00	2. 06	0. 26	1809. 50	293. 80	237. 68	1. 07	0. 84	33. 40	100. 00	14397. 40	7. 42	93. 79		
2008	81. 00	3. 01	0. 23	1762. 10	297. 60	304. 73	1. 26	0. 84	30. 90	100. 00	13271. 60	8. 83	93. 79		
2009	110. 00	4. 13	0. 25	1592. 20	265. 90	274. 23	1. 40	0. 85	31. 40	100. 00	11046. 40	7. 67	93. 79		
2010	113. 00	4. 11	0. 29	1687. 10	269. 60	308. 33	1. 58	0. 85	31. 30	100. 00	13992. 50	7. 67	93. 83		
2011	185. 00	6. 55	0. 24	1681. 10	304. 50	534. 40	2. 31	0. 86	31. 90	100. 00	13315. 20	7. 28	93. 83		
2012	274. 00	9. 37	0. 24	1710. 20	317. 50	491. 08	2. 12	0. 86	32. 50	100. 00	12405. 20	6. 95	93. 83		
2013	255. 00	8. 76	0. 24	1675. 50	320. 40	435. 76	1. 73	0. 86	32. 90	100. 00	14290. 00	7. 38	93. 83		
2014	273. 00	9. 01	0. 23	1689. 00	334. 70	384. 36	1. 45	0. 87	35. 60	100. 00	14129. 50	7. 34	94. 65		
2015	300. 00	10. 50	0. 24	1747. 90	331. 80	340. 12	1. 50	0. 85	34. 80	100. 00	11830. 00	7. 29	94. 65		
2016								0. 87		100. 00					

7. France

Year	id	along BR	OECD	income group	GDP per capita	Population	Forest area	Biodiver-ity & Habitat	Water resources	Natural resources	Natural resources	Renewable power generation	Renewable power generation
Year	id	brc	oecd	inc_ group	gdppc	pop	forest	bio	water	natural	natural_ pct	renew_ g	renew_ g_ pct
2005	265	0	1	high	34760	63. 18	28. 97			945. 60	0. 04	56803. 44	9. 86
2006	266	0	1	high	36444	63. 62	29. 17		3139. 48	1140. 86	0. 05	62934. 33	10. 95
2007	267	0	1	high	41508	64. 02	29. 38	98. 45	3124. 21	1368. 28	0. 05	66570. 12	11. 69
2008	268	0	1	high	45334	64. 37	29. 58	98. 67	3108. 57	1602. 37	0. 05	74437. 84	12. 98
2009	269	0	1	high	41575	64. 71	29. 79	99	3092. 93	1371. 18	0. 05	70313. 68	13. 13
2010	270	0	1	high	40638	65. 03	30. 00	99. 4	3077. 29	1464. 80	0. 06	78861. 82	13. 86
2011	271	0	1	high	43791	65. 34	30. 20	99. 33	3061. 65	1568. 14	0. 05	64986. 30	11. 57
2012	272	0	1	high	40875	65. 66	30. 41	99. 34	3046. 00	1350. 30	0. 05	83759. 28	14. 83
2013	273	0	1	high	42593	66. 00	30. 61	99. 5	3030. 93	1343. 32	0. 05	97550. 00	17. 05
2014	274	0	1	high	43009	66. 32	30. 82	99. 62	3015. 86	1403. 85	0. 05	92760. 95	16. 46
2015	275	0	1	high	36613	66. 59	31. 03	99. 62	2999. 87	1016. 68	0. 04	90143. 54	15. 86
2016	276	0	1	high	36870	66. 86				972. 94	0. 04		

Year	Renewable capacity	Renewable capacity	Energy efficiency	Green transportation	Green Building	R&D Technology Competitiveness	R&D Technology Competitiveness	HDI	Inequality	Access to electricity	CO_2	PM2. 5 concentration	Air pollution-average exposure to PM2. 5	Missing data 15-in total	Percentage of overall missing data
	renew_ c	renew_ c_ pct	ee	trans	bldg	rd	rd_ pct	hdi	inequality	electricity	co2	pm_ con	pm_ expo		
2005	25740. 00	22. 24	0. 12			44949. 43	2. 05	0. 87	29. 93	100. 00		12. 26		0	11. 3%
2006	26428. 00	22. 84	0. 11	131. 20	1345. 30	47505. 43	2. 05	0. 88	29. 92	100. 00	5701. 70				
2007	27260. 00	23. 39	0. 10	130. 40	1213. 90	53557. 20	2. 02	0. 88	32. 63	100. 00	5540. 50		81. 42		
2008	28362. 00	24. 09	0. 09	124. 40	1304. 80	59923. 74	2. 05	0. 88	33. 08	100. 00	5442. 90		80. 83		
2009	29612. 00	24. 88	0. 09	123. 00	1327. 90	59399. 84	2. 21	0. 88	33. 78	100. 00	5159. 70		81. 44		
2010	31159. 00	25. 02	0. 10	123. 40	1297. 50	57435. 53	2. 17	0. 88	33. 35	100. 00	5245. 40	12. 24	81. 34		
2011	32426. 00	25. 48	0. 09	122. 90	1062. 40	62661. 41	2. 19	0. 89	33. 10	100. 00	4751. 50	12. 21	83. 56		
2012	33587. 00	25. 99	0. 09	121. 90	1154. 00	59736. 31	2. 23	0. 89	34. 76	100. 00	4777. 20	12. 26	84. 83		
2013	34739. 00	27. 05	0. 09	121. 00	1178. 20	62801. 16	2. 23	0. 89	35. 38	100. 00	4795. 50	12. 29	84. 49		
2014	36050. 00	27. 96	0. 09	121. 30	975. 40	63865. 41	2. 24	0. 90	36. 01	100. 00	4288. 90	12. 33	85. 44		
2015	38049. 00	29. 42	0. 08	122. 40	984. 30	54404. 95	2. 23	0. 88	36. 64	100. 00	4368. 00	11. 89	87. 80		
2016								0. 90		100. 00		11. 87			

8. Germany

Year	id	along BR	OECD	income group	GDP per capita	Population	Forest area	Biodiver-ity & Habitat	Water resources	Natural resources	Natural resources	Renewable power generation	Renewable power generation
Year	id	brc	oecd	inc_ group	gdppc	pop	forest	bio	water	natural	natural_ pct	renew_ g	renew_ g_ pct
2005	277	0	1	high	34697	82. 47	32. 64					63193. 08	10. 15
2006	278	0	1	highincome	36448	82. 38	32. 66		1301. 47			72403. 37	11. 32
2007	279	0	1	high	41815	82. 27	32. 68	100	1300. 65			89283. 81	13. 94
2008	280	0	1	high	45699	82. 11	32. 70	100	1306. 61			94134. 78	14. 70
2009	281	0	1	high	41733	81. 90	32. 71	100	1312. 56			95766. 60	16. 08
2010	282	0	1	high	41786	81. 78	32. 73	100	1318. 51			105879. 53	16. 73
2011	283	0	1	high	46810	80. 27	32. 74	100	1324. 47			124963. 38	20. 38
2012	284	0	1	high	44065	80. 43	32. 75	100	1330. 42			144859. 65	23. 00
2013	285	0	1	high	46531	80. 65	32. 72	100	1325. 85			153758. 36	24. 07
2014	286	0	1	high	48043	80. 98	32. 72	100	1321. 27			164043. 44	26. 13
2015	287	0	1	high	41324	81. 69	32. 73	100	1333. 62			189096. 81	29. 23
2016	288	0	1	high	42233	82. 35							

Year	Renewable capacity	Renewable capacity	Energy efficiency	Green transportation	Green Building	R&D Tec-hnology Competi-tiveness	R&D Technology Competi-tiveness	HDI	Inequality	Access to electricity	CO_2	PM2. 5 concen-tration	Air pollution-average exposure to PM2. 5	Missing data 15-in total	Percentage of overall missing data
	renew_ c	renew_ c_ pct	ee	trans	bldg	rd	rd_ pct	hdi	inequality	electricity	co2	pm_ con	pm_ expo		
2005	31198. 00	24. 26	0. 12			69344. 56	2. 42	0. 91		100. 00		13. 39		1	20. 0%
2006	34194. 00	25. 89	0. 12	153. 40	2071. 50	73841. 57	2. 46	0. 92	32. 78	100. 00	9846. 30	13. 45			
2007	37078. 00	27. 20	0. 10	148. 30	1597. 60	84260. 97	2. 45	0. 92	32. 40	100. 00	9467. 90	13. 51	68. 27		
2008	40631. 00	28. 32	0. 09	148. 90	1900. 60	97519. 85	2. 60	0. 92	31. 29	100. 00	9599. 10	13. 58	67. 88		
2009	47391. 00	31. 24	0. 09	147. 00	1771. 70	93200. 45	2. 73	0. 92	31. 51	100. 00	8949. 80	13. 64	68. 78		
2010	55831. 00	34. 32	0. 10	148. 10	1885. 10	92730. 72	2. 71	0. 93	31. 14	100. 00	9452. 40	13. 70	72. 92		
2011	65474. 00	37. 22	0. 08	149. 90	1626. 10	105172. 34	2. 80	0. 93	30. 13	100. 00	9110. 10	13. 68	74. 06		
2012	75144. 00	42. 38	0. 09	148. 50	1715. 10	101581. 92	2. 87	0. 93	29. 86	100. 00	9260. 40	13. 30	74. 95		
2013	82219. 00	44. 18	0. 08	152. 30	1839. 20	105719. 56	2. 82	0. 93	29. 38	100. 00	9471. 80	13. 66	71. 85		
2014	88643. 00	44. 68	0. 08	154. 50	1578. 40	112526. 86	2. 89	0. 93	28. 90	100. 00	8931. 10	13. 87	75. 49		
2015	95840. 00	46. 97	0. 07	157. 50	1609. 60	97132. 87	2. 88	0. 93	28. 42	100. 00	8933. 70	13. 48	71. 40		
2016								0. 94		100. 00		13. 46			

9. Greece

Year	id	along BR	OECD	income group	GDP per capita	Population	Forest area	Biodiver-ity & Habitat	Water resources	Natural resources	Natural resources	Renewable power generation	Renewable power generation
Year	id	brc	oecd	inc_ group	gdppc	pop	forest	bio	water	natural	natural_ pct	renew_ g	renew_ g_ pct
2005	109	1	1	high	22552	10. 99	29. 11			439. 00	0. 18	6469. 92	10. 78
2006	110	1	1	high	24801	11. 02	29. 34		5225. 83	673. 28	0. 25	7756. 84	12. 76
2007	111	1	1	high	28827	11. 05	29. 58	94. 27	5249. 59	1167. 15	0. 37	4651. 51	7. 33
2008	112	1	1	high	31997	11. 08	29. 81	94. 39	5249. 92	1453. 95	0. 41	5826. 50	9. 14
2009	113	1	1	high	29711	11. 11	30. 04	94. 27	5250. 25	446. 60	0. 14	8221. 31	13. 40
2010	114	1	1	high	26919	11. 12	30. 28	94. 15	5250. 58	958. 85	0. 32	10526. 59	18. 34
2011	115	1	1	high	25915	11. 10	30. 51	94. 08	5250. 91	1277. 70	0. 44	8179. 33	13. 76
2012	116	1	1	high	22243	11. 05	30. 75	94. 15	5251. 24	712. 06	0. 29	10182. 40	16. 70
2013	117	1	1	high	21843	10. 97	30. 98	93. 96	5288. 02	494. 00	0. 21	14358. 55	25. 12
2014	118	1	1	high	21627	10. 89	31. 22	93. 76	5324. 81	464. 91	0. 20	12208. 69	24. 19
2015	119	1	1	high	18036	10. 82	31. 45	93. 69	5303. 00	177. 57	0. 09	14864. 90	28. 66
2016	120	1	1	high	17891	10. 78	31. 45			169. 40	0. 09		

Year	Renewable capacity	Renewable capacity	Energy efficiency	Green transportation	Green Building	R&D Tec-hnology Competi-tiveness	R&D Technology Competi-tiveness	HDI	Inequality	Access to electricity	CO_2	PM2. 5 concen-tration	Air pollution-average exposure to PM2. 5	Missing data 15-in total	Percentage of overall missing data
	renew_ c	renew_ c_ pct	ee	trans	bldg	rd	rd_ pct	hdi	inequality	electricity	co2	pm_ con	pm_ expo		
2005	3598. 00	27. 04	0. 12			1436. 30	0. 58	0. 85	33. 20	100. 00		12. 70		0	9. 6%
2006	3888. 00	28. 65	0. 11	2022. 10	1004. 70	1532. 85	0. 56	0. 85	34. 30	100. 00	8543. 90	12. 20			
2007	4005. 00	29. 26	0. 09	2070. 40	914. 30	1833. 88	0. 58	0. 86	34. 30	100. 00	8866. 40	12. 00	75. 82		
2008	4210. 00	29. 54	0. 09	2003. 40	894. 50	2345. 86	0. 66	0. 86	33. 40	100. 00	8521. 70	11. 70	75. 94		
2009	4418. 00	30. 20	0. 09	2232. 20	778. 50	2060. 33	0. 62	0. 86	33. 10	100. 00	8122. 00	11. 40	76. 69		
2010	4715. 00	30. 79	0. 09	1976. 70	699. 80	1791. 67	0. 60	0. 85	32. 90	100. 00	7501. 00	11. 30	76. 28		
2011	5476. 00	33. 14	0. 09	1768. 20	891. 70	1934. 06	0. 67	0. 85	33. 50	100. 00	7399. 10	11. 60	77. 94		
2012	6525. 00	36. 76	0. 11	1485. 90	675. 50	1720. 46	0. 70	0. 86	34. 30	100. 00	6975. 00	11. 60	78. 81		
2013	7626. 00	40. 45	0. 10	1486. 80	405. 30	1939. 45	0. 81	0. 86	34. 40	100. 00	6282. 90	11. 11	77. 55		
2014	7963. 00	42. 14	0. 10	1500. 50	403. 10	1970. 28	0. 84	0. 87	34. 50	100. 00	6038. 40	10. 95	81. 08		
2015	8087. 00	42. 69	0. 12	1536. 30	560. 90	1867. 16	0. 96	0. 85	34. 20	100. 00	5947. 50	10. 80	82. 79		
2016									0. 87		100. 00				

10. India

Year	id	along BR	OECD	income group	GDP per capita	Population	Forest area	Biodiver-ity & Habitat	Water resources	Natural resources	Natural resources	Renewable power generation	Renewable power generation
Year	id	brc	oecd	inc_ group	gdppc	pop	forest	bio	water	natural	natural_ pct	renew_ g	renew_ g_ pct
2005	121	1	0	lower middle	729	1144. 12	22. 77			31871. 27	3. 82	118935. 00	16. 62
2006	122	1	0	lower middle	817	1161. 98	22. 91		1240. 29	41054. 59	4. 33	135725. 00	17. 54
2007	123	1	0	lower middle	1050	1179. 68	23. 05	64. 78	1225. 75	60419. 09	4. 88	147122. 00	17. 86
2008	124	1	0	lower middle	1023	1197. 15	23. 19	64. 65	1209. 57	92109. 49	7. 52	140062. 00	16. 51
2009	125	1	0	lower middle	1125	1214. 27	23. 33	63. 45	1193. 39	49086. 33	3. 59	143650. 00	15. 66
2010	126	1	0	lower middle	1388	1230. 98	23. 47	62. 2	1177. 20	79803. 73	4. 67	157135. 00	16. 04
2011	127	1	0	lower middle	1456	1247. 24	23. 53	62. 14	1161. 02	95033. 44	5. 23	186077. 00	17. 32
2012	128	1	0	lower middle	1444	1263. 07	23. 59	62. 24	1144. 83	73702. 86	4. 04	176689. 00	15. 73
2013	129	1	0	lower middle	1456	1278. 56	23. 65	61. 29	1131. 21	68258. 86	3. 67	206622. 00	17. 35
2014	130	1	0	lower middle	1577	1293. 86	23. 71	61. 59	1117. 59	55992. 13	2. 74	210273. 00	16. 25
2015	131	1	0	lower middle	1582	1309. 05	23. 77	61. 28	1099. 85	37845. 28	1. 83	212195. 00	15. 34
2016	132	1	0	lower middle	1710	1324. 17	23. 77			43194. 35	1. 91		

Year	Renewable capacity	Renewable capacity	Energy efficiency	Green transportation	Green Building	R&D Technology Competitiveness	R&D Technology Competitiveness	HDI	Inequality	Access to electricity	CO_2	PM2. 5 concentration	Air pollution-average exposure to PM2. 5	Missing data 15-in total	Percentage of overall missing data
	renew_ c	renew_ c_ pct	ee	trans	bldg	rd	rd_ pct	hdi	inequality	electricity	co2	pm_ con	pm_ expo		
2005	36756. 00	24. 47	0. 62			6765. 75	0. 81	0. 55	48. 00	66. 93		27. 30		0	9. 6%
2006	40984. 00	25. 46	0. 57	103. 00	73. 90	7573. 87	0. 80	0. 56	45. 76	67. 90	990. 20	28. 30			
2007	43754. 00	24. 88	0. 46	120. 10	74. 30	10093. 02	0. 81	0. 56	43. 52	69. 99	1072. 10	29. 30	25. 46		
2008	46533. 00	25. 21	0. 49	135. 30	74. 20	10617. 26	0. 87	0. 57	41. 28	71. 55	1124. 00	30. 60	25. 08		
2009	47789. 00	24. 04	0. 49	146. 00	78. 10	11532. 70	0. 84	0. 58	39. 04	75. 00	1246. 20	31. 00	24. 67		
2010	50632. 00	23. 06	0. 41	155. 30	80. 40	14041. 98	0. 82	0. 59	36. 80	76. 30	1295. 10	31. 10	15. 05		
2011	55074. 00	21. 81	0. 40	166. 10	82. 90	15085. 83	0. 83	0. 60	50. 30	67. 60	1341. 80	31. 70	15. 11		
2012	57912. 00	20. 81	0. 42	173. 50	80. 80	14229. 57	0. 78	0. 61	35. 90	79. 90	1417. 20	32. 00	14. 93		
2013	62006. 00	21. 03	0. 42	174. 80	82. 20	13575. 62	0. 73	0. 62	38. 05	79. 60	1447. 90	33. 15	14. 77		
2014	67197. 00	21. 24	0. 40	179. 70	85. 10	13837. 88	0. 68	0. 63	37. 05	81. 24	1558. 60	33. 81	13. 77		
2015	68859. 00	23. 98	0. 35	194. 10	86. 60	12989. 69	0. 63	0. 59	36. 04	88. 00	1575. 80	34. 47	14. 06		
2016									0. 64		84. 53				

11. Indonesia

Year	id	along BR	OECD	income group	GDP per capita	Population	Forest area	Biodiver-ity & Habitat	Water resources	Natural resources	Natural resources	Renewable power generation	Renewable power generation
Year	id	brc	oecd	inc_ group	gdppc	pop	forest	bio	water	natural	natural_ pct	renew_ g	renew_ g_ pct
2005	37	1	0	lower middle	1264	226. 71	54. 02			25055. 19	8. 75	17351. 00	13. 61
2006	38	1	0	lower middle	1590	229. 84	53. 64		8769. 02	31973. 39	8. 75	16313. 00	12. 26
2007	39	1	0	lower middle	1861	232. 99	53. 26	80. 98	8665. 64	39968. 54	9. 22	18343. 00	12. 90
2008	40	1	0	lower middle	2168	236. 16	52. 88	80. 97	8554. 96	53659. 78	10. 48	19884. 00	13. 32
2009	41	1	0	lower middle	2263	239. 34	52. 51	81. 18	8444. 28	31374. 90	5. 79	20746. 00	13. 23
2010	42	1	0	lower middle	3125	242. 52	52. 13	81. 07	8333. 60	51325. 43	6. 77	26913. 00	15. 85
2011	43	1	0	lower middle	3648	245. 71	51. 75	81. 41	8222. 92	75974. 27	8. 48	21997. 00	11. 99
2012	44	1	0	lower middle	3701	248. 88	51. 37	80. 83	8112. 24	59229. 98	6. 43	22477. 00	11. 24
2013	45	1	0	lower middle	3632	252. 03	50. 99	80. 89	8012. 91	51650. 08	5. 64	26506. 00	12. 27
2014	46	1	0	lower middle	3500	255. 13	50. 62	81. 17	7913. 58	39771. 05	4. 45	26149. 00	11. 48
2015	47	1	0	lower middle	3347	258. 16	50. 24	81. 41	7796. 01	22394. 98	2. 59	24922. 00	10. 65
2016	48	1	0	lower middle	3570	261. 12	50. 24			23272. 13	2. 50		

Year	Renewable capacity	Renewable capacity	Energy efficiency	Green transportation	Green Building	R&D Tec-hnology Competi-tiveness	R&D Technology Competi-tiveness	HDI	Inequality	Access to electricity	CO_2	PM2. 5 concen-tration	Air pollution-average exposure to PM2. 5	Missing data 15-in total	Percentage of overall missing data
	renew_ c	renew_ c_ pct	ee	trans	bldg	rd	rd_ pct	hdi	inequality	electricity	co2	pm_ con	pm_ expo		
2005	4041. 00	15. 15	0. 63					0. 64	36. 00	86. 19		8. 72	93. 49	0	12. 1%
2006	4349. 00	16. 11	0. 50	292. 80	122. 90			0. 64	33. 00	90. 62	1483. 20	9. 83	93. 05		
2007	4434. 10	16. 22	0. 42	304. 00	122. 40			0. 65	36. 00	91. 10	1523. 20	9. 57	92. 13		
2008	4743. 26	15. 24	0. 36	330. 00	109. 50			0. 66	35. 00	92. 73	1487. 70	9. 34	92. 13		
2009	4612. 26	12. 54	0. 37	372. 60	92. 80	450. 95	0. 08	0. 66	37. 00	93. 55	1530. 50	8. 56	92. 13		
2010	4820. 20	12. 52	0. 28	419. 30	84. 90	633. 77	0. 08	0. 67	38. 00	94. 15	1556. 90	8. 30	90. 93		
2011	5050. 30	10. 58	0. 23	449. 00	81. 60	752. 53	0. 08	0. 68	41. 00	94. 83	1571. 20	8. 63	90. 93		
2012	5963. 20	11. 22	0. 23	491. 50	83. 50	776. 51	0. 08	0. 68	39. 00	96. 00	1584. 20	9. 21	90. 93		
2013	5988. 13	10. 77	0. 24	511. 60	87. 30	774. 90	0. 08	0. 68	39. 47	96. 46	1584. 00	8. 64	90. 93		
2014	6336. 83	10. 40	0. 25	502. 90	87. 70	759. 00	0. 09	0. 69	40. 90	97. 01	1709. 00	8. 55	88. 66		
2015	6232. 00	8. 46	0. 13	499. 30	86. 80	737. 42	0. 09	0. 67	41. 65	97. 54	1715. 70	8. 47	88. 66		
2016								0. 69		97. 62					

12. Japan

Year	id	along BR	OECD	income group	GDP per capita	Population	Forest area	Biodiver-ity & Habitat	Water resources	Natural resources	Natural resources	Renewable power generation	Renewable power generation
Year	id	brc	oecd	inc_ group	gdppc	pop	forest	bio	water	natural	natural_ pct	renew_ g	renew_ g_ pct
2005	289	0	1	high	37218	127. 77	68. 41			579. 99	0. 01	95777. 62	8. 41
2006	290	0	1	high	35434	127. 85	68. 43		3355. 65	723. 16	0. 02	107738. 52	9. 45
2007	291	0	1	high	35275	128. 00	68. 44	92. 98	3359. 35	874. 90	0. 02	95437. 63	8. 20
2008	292	0	1	high	39339	128. 06	68. 46	93. 16	3361. 31	1077. 60	0. 02	97402. 17	8. 79
2009	293	0	1	high	40855	128. 05	68. 48	93. 02	3363. 27	947. 41	0. 02	98344. 35	9. 15
2010	294	0	1	high	44508	128. 07	68. 48	92. 95	3365. 22	1006. 03	0. 02	120961. 04	10. 53
2011	295	0	1	high	48168	127. 83	68. 48	92. 66	3367. 18	1246. 83	0. 02	124131. 72	11. 46
2012	296	0	1	high	48603	127. 63	68. 47	92. 75	3369. 14	1289. 09	0. 02	119195. 34	11. 19
2013	297	0	1	high	40454	127. 45	68. 47	92. 69	3373. 81	1235. 87	0. 02	130155. 77	12. 21
2014	298	0	1	high	38109	127. 28	68. 46	92. 39	3378. 48	1100. 92	0. 02	148940. 49	14. 06
2015	299	0	1	high	34568	127. 14	68. 46	92. 31	3378. 80	894. 20	0. 02	166456. 40	15. 98
2016	300	0	1	high	38972	126. 99				835. 00	0. 02		

Year	Renewable capacity	Renewable capacity	Energy efficiency	Green transpor-tation	Green Building	R&D Tec-hnology Competi-tiveness	R&D Technology Competi-tiveness	HDI	Inequality	Access to electricity	CO_2	PM2. 5 concen-tration	Air pollution-average exposure to PM2. 5	Missing data 15-in total	Percentage of overall missing data
	renew_ c	renew_ c_ pct	ee	trans	bldg	rd	rd_ pct	hdi	inequality	electricity	co2	pm_ con	pm_ expo		
2005	46398. 00	16. 70	0. 11			151307. 66	3. 18	0. 88		100. 00		13. 21		0	13. 3%
2006	46463. 00	16. 62	0. 11	228. 70	1087. 40	148491. 72	3. 28	0. 88		100. 00	9092. 30	13. 03			
2007	46385. 00	16. 56	0. 11	225. 80	1053. 80	150703. 27	3. 34	0. 88		100. 00	9426. 80	12. 86	78. 15		
2008	46372. 00	16. 47	0. 10	217. 30	958. 00	168032. 38	3. 34	0. 88	32. 11	100. 00	8751. 60	12. 69	76. 53		
2009	46352. 00	16. 23	0. 09	213. 70	1016. 40	168874. 79	3. 23	0. 89	31. 10	100. 00	8278. 60	12. 51	76. 45		
2010	43967. 00	15. 25	0. 09	214. 20	951. 50	178943. 18	3. 14	0. 89		100. 00	8683. 20	12. 34	72. 73		
2011	44816. 00	15. 27	0. 07	211. 70	957. 30	199961. 65	3. 25	0. 90		100. 00	9119. 20	11. 79	73. 12		
2012	45277. 00	15. 26	0. 07	216. 60	920. 90	199049. 32	3. 21	0. 90		100. 00	9477. 00	11. 97	78. 43		
2013	45302. 00	14. 89	0. 09	215. 80	979. 90	170969. 76	3. 32	0. 90		100. 00	9656. 60	12. 40	78. 32		
2014	45987. 00	14. 50	0. 09	208. 20	982. 40	164884. 96	3. 40	0. 91		100. 00	9316. 80	12. 88	80. 37		
2015	46396. 00	14. 32	0. 07	207. 80	921. 40	144314. 81	3. 28	0. 89		100. 00	8990. 10	13. 15	82. 04		
2016								0. 91		100. 00		13. 16			

13. Kazakhstan

Year	id	along BR	OECD	income group	GDP per capita	Population	Forest area	Biodiver-ity & Habitat	Water resources	Natural resources	Natural resources	Renewable power generation	Renewable power generation
Year	id	brc	oecd	inc_ group	gdppc	pop	forest	bio	water	natural	natural_ pct	renew_ g	renew_ g_ pct
2005	133	1	0	upper middle	3771	15. 15	1. 24			18517. 39	32. 42	7856. 00	11. 58
2006	134	1	0	upper middle	5292	15. 31	1. 23		4214. 33	24788. 83	30. 60	7768. 00	10. 84
2007	135	1	0	upper middle	6771	15. 48	1. 23	51. 33	4155. 85	28327. 43	27. 02	8171. 00	10. 67
2008	136	1	0	upper middle	8514	15. 67	1. 23	50. 61	4091. 11	41149. 39	30. 84	7460. 00	9. 29
2009	137	1	0	upper middle	7165	16. 09	1. 23	48. 74	4026. 38	22270. 78	19. 31	6879. 00	8. 74
2010	138	1	0	upper middle	9071	16. 32	1. 23	48. 38	3961. 64	34062. 99	23. 01	8022. 00	9. 71
2011	139	1	0	upper middle	12103	16. 56	1. 23	50. 94	3896. 90	53014. 51	26. 46	7883. 00	9. 10
2012	140	1	0	upper middle	12858	16. 79	1. 23	50. 32	3832. 16	47531. 26	22. 01	7640. 00	8. 23
2013	141	1	0	upper middle	14310	17. 04	1. 23	49. 99	3777. 17	43946. 60	18. 03	7737. 00	7. 51
2014	142	1	0	upper middle	13155	17. 29	1. 23	51. 66	3722. 17	39826. 60	17. 51	8277. 00	7. 88
2015	143	1	0	upper middle	10508	17. 54	1. 23	51. 5	3651. 52	18989. 72	10. 30	9448. 00	8. 87
2016	144	1	0	upper middle	7715	17. 79	1. 23			20647. 45	15. 04		

Year	Renewable capacity	Renewable capacity	Energy efficiency	Green transportation	Green Building	R&D Tec-hnology Competi-tiveness	R&D Technology Competi-tiveness	HDI	Inequality	Access to electricity	CO_2	PM2.5 concen-tration	Air pollution-average exposure to PM2.5	Missing data 15-in total	Percentage of overall missing data
	renew_ c	renew_ c_ pct	ee	trans	bldg	rd	rd_ pct	hdi	inequality	electricity	co2	pm_ con	pm_ expo		
2005	2217. 00	11. 83	0. 89			162. 02	0. 28	0. 75	42. 00	99. 22		8. 69	89. 01	0	8. 8%
2006	2217. 00	11. 83	0. 76	688. 10	85. 30	196. 95	0. 24	0. 76	41. 40	99. 77	11296. 40	8. 93	88. 88		
2007	2217. 00	11. 83	0. 63	781. 30	57. 70	219. 82	0. 21	0. 76	30. 90	99. 33	12089. 30	8. 38	88. 99		
2008	2217. 00	11. 83	0. 52	879. 40	1245. 60	288. 11	0. 22	0. 76	28. 80	99. 42	14689. 40	7. 71	88. 99		
2009	2217. 00	11. 83	0. 55	772. 90	906. 00	264. 46	0. 23	0. 77	26. 70	100. 00	12595. 70	7. 63	88. 99		
2010	2217. 00	11. 54	0. 47	805. 90	1013. 40	227. 29	0. 15	0. 77	27. 80	99. 64	13545. 30	7. 70	88. 39		
2011	2217. 00	10. 97	0. 39	813. 20	1359. 00	308. 05	0. 15	0. 78	28. 90	99. 80	14182. 10	7. 71	88. 39		
2012	2219. 00	10. 46	0. 34	868. 60	984. 80	356. 97	0. 17	0. 79	27. 46	99. 87	13926. 10	7. 35	88. 39		
2013	2221. 00	9. 99	0. 33	806. 20	1058. 30	417. 62	0. 17	0. 79	26. 33	99. 95	14625. 80	7. 07	88. 39		
2014	2510. 00	10. 04	0. 34	790. 30	1369. 00	380. 07	0. 17	0. 80	22. 02	99. 99	13291. 50	6. 86	88. 83		
2015	2600. 00	10. 36	0. 19	844. 50	1246. 20	312. 36	0. 17	0. 77	20. 19	100. 00	12833. 20	6. 65	88. 83		
2016								0. 80		100. 00					

14. Kyrgyzstan

Year	id	along BR	OECD	income group	GDP per capita	Population	Forest area	Biodiver-ity & Habitat	Water resources	Natural resources	Natural resources	Renewable power generation	Renewable power generation
Year	id	brc	oecd	inc_ group	gdppc	pop	forest	bio	water	natural	natural_ pct	renew_ g	renew_ g_ pct
2005	145	1	0	lower middle	477	5. 16	4. 53					12788. 00	85. 88
2006	146	1	0	lower middle	543	5. 22	4. 33		9437. 27			12472. 00	85. 88
2007	147	1	0	lower middle	722	5. 27	4. 13	82. 4	9287. 45			12736. 00	85. 88
2008	148	1	0	lower middle	966	5. 32	3. 93	81. 91	9175. 22			10124. 00	85. 88
2009	149	1	0	lower middle	871	5. 38	3. 73	81. 63	9062. 98			10217. 00	92. 19
2010	150	1	0	lower middle	880	5. 45	3. 53	81. 44	8950. 75			11108. 00	91. 80
2011	151	1	0	lower middle	1124	5. 51	3. 49	80. 8	8838. 51			14139. 00	93. 28
2012	152	1	0	lower middle	1178	5. 61	3. 45	80. 2	8726. 28			14179. 00	93. 48
2013	153	1	0	lower middle	1282	5. 72	3. 40	79. 73	8555. 58			13097. 00	93. 48
2014	154	1	0	lower middle	1280	5. 84	3. 36	79. 99	8384. 89			13298. 00	91. 26
2015	155	1	0	lower middle	1103	5. 96	3. 32	79. 9	8308. 15			11100. 00	85. 19
2016	156	1	0	lower middle	1078	6. 08	3. 32						

Year	Renewable capacity	Renewable capacity	Energy efficiency	Green transportation	Green Building	R&D Technology Competitiveness	R&D Technology Competitiveness	HDI	Inequality	Access to electricity	CO_2	PM2.5 concentration	Air pollution-average exposure to PM2.5	Missing data 15-in total	Percentage of overall missing data
	renew_ c	renew_ c_ pct	ee	trans	bldg	rd	rd_ pct	hdi	inequality	electricity	co2	pm_ con	pm_ expo		
2005	2940.00	78.55						0.62	47.50	99.44		9.14	84.50	4	32.5%
2006	2943.00	78.94		222.90	46.20			0.63	46.00	99.46	934.70	8.76	83.94		
2007	2945.00	78.76		331.40	49.00			0.63	33.91	99.45	1132.00	8.79	83.10		
2008	2944.00	78.70		368.20	44.20			0.64	31.45	99.45	1317.30	8.18	83.10		
2009	2944.00	78.72		421.90	41.60			0.64	29.87	99.48	1211.70	8.00	83.10		
2010	3064.00	79.40		363.90	33.70			0.64	30.13	99.00	1108.90	7.84	80.83		
2011	3064.00	79.40		502.50	44.40			0.65	27.84	99.58	1307.80	7.83	80.83		
2012	3064.00	79.40		701.70	48.40			0.66	27.36	99.80	1706.00	7.80	80.83		
2013	3064.00	79.36		688.70	83.60			0.66	28.82	99.73	1551.80	7.38	80.83		
2014	3064.00	79.30		540.40	259.40			0.67	26.82	99.80	1548.10	7.18	79.03		
2015	3064.00	76.22		449.50	305.40			0.64	21.31	99.91	1659.00	6.97	79.03		
2016								0.67		100.00					

15. Latvia

Year	id	along BR	OECD	income group	GDP per capita	Population	Forest area	Biodiver-ity & Habitat	Water resources	Natural resources	Natural resources	Renewable power generation	Renewable power generation
Year	id	brc	oecd	inc_ group	gdppc	pop	forest	bio	water	natural	natural_ pct	renew_ g	renew_ g_ pct
2005	193	1	1	high	7550	2. 24	53. 01			221. 03	1. 31	3414. 00	69. 59
2006	194	1	1	high	9652	2. 22	53. 19		7599. 57	250. 00	1. 17	2786. 00	56. 96
2007	195	1	1	high	14019	2. 20	53. 38	96. 66	7698. 86	295. 43	0. 96	2828. 00	59. 27
2008	196	1	1	high	16324	2. 18	53. 56	96. 76	7824. 51	235. 20	0. 66	3212. 00	60. 90
2009	197	1	1	high	12208	2. 14	53. 76	96. 88	7950. 16	221. 23	0. 85	3556. 00	63. 85
2010	198	1	1	high	11320	2. 10	53. 89	97. 02	8075. 81	291. 92	1. 23	3635. 00	54. 85
2011	199	1	1	high	13781	2. 06	53. 93	96. 89	8201. 46	341. 47	1. 20	3077. 00	50. 49
2012	200	1	1	high	13775	2. 03	53. 95	98. 17	8327. 11	362. 82	1. 29	4109. 00	66. 63
2013	201	1	1	high	15016	2. 01	53. 95	98. 17	8411. 76	318. 09	1. 05	3534. 00	56. 92
2014	202	1	1	high	15692	1. 99	53. 97	98. 22	8496. 42	328. 08	1. 05	2804. 00	54. 54
2015	203	1	1	high	13665	1. 98	53. 97	98. 16	8646. 96	254. 42	0. 94	2776. 00	50. 17
2016	204	1	1	high	14071	1. 96	53. 97			272. 02	0. 99		

Year	Renewable capacity	Renewable capacity	Energy efficiency	Green transportation	Green Building	R&D Technology Competitiveness	R&D Technology Competitiveness	HDI	Inequality	Access to electricity	CO_2	PM2. 5 concentration	Air pollution-average exposure to PM2. 5	Missing data 15-in total	Percentage of overall missing data
	renew_ c	renew_ c_ pct	ee	trans	bldg	rd	rd_ pct	hdi	inequality	electricity	co2	pm_ con	pm_ expo		
2005	1556. 00	71. 84	0. 27			89. 52	0. 53	0. 81	36. 20	100. 00		10. 20		0	9. 6%
2006	1557. 00	72. 38	0. 22	1493. 90	419. 30	139. 31	0. 65	0. 82	38. 90	100. 00	3618. 00	10. 50			
2007	1557. 00	73. 00	0. 16	1703. 90	423. 00	170. 83	0. 55	0. 82	35. 40	100. 00	3797. 30	10. 90	90. 08		
2008	1557. 00	72. 25	0. 13	1627. 50	414. 10	206. 64	0. 58	0. 82	37. 50	100. 00	3643. 70	10. 10	89. 19		
2009	1557. 00	62. 23	0. 17	1451. 20	418. 50	118. 36	0. 45	0. 82	37. 50	100. 00	3354. 00	9. 34	87. 60		
2010	1599. 00	62. 53	0. 19	1516. 40	483. 70	145. 02	0. 61	0. 82	35. 90	100. 00	3855. 60	9. 61	89. 42		
2011	1606. 00	62. 32	0. 15	1370. 20	444. 60	198. 06	0. 70	0. 82	35. 10	100. 00	3563. 30	9. 28	89. 93		
2012	1632. 00	61. 35	0. 16	1340. 70	424. 20	186. 46	0. 67	0. 83	35. 70	100. 00	3438. 20	8. 85	89. 93		
2013	1654. 00	56. 82	0. 14	1371. 60	408. 30	185. 12	0. 61	0. 84	35. 20	100. 00	3433. 20	8. 77	89. 93		
2014	1657. 00	56. 67	0. 14	1445. 80	424. 10	215. 75	0. 69	0. 84	35. 50	100. 00	3371. 70	8. 53	90. 66		
2015	1658. 00	56. 49	0. 12	1546. 00	408. 30	169. 03	0. 63	0. 82	35. 40	100. 00	3461. 80	8. 29	90. 66		
2016								0. 85		100. 00					

16. Malaysia

Year	id	along BR	OECD	income group	GDP per capita	Population	Forest area	Biodiver-ity & Habitat	Water resources	Natural resources	Natural resources	Renewable power generation	Renewable power generation
Year	id	brc	oecd	inc_ group	gdppc	pop	forest	bio	water	natural	natural_ pct	renew_ g	renew_ g_ pct
2005	25	1	0	upper middle	5564	25. 66	63. 58			16127. 02	11. 30	5192. 00	6. 28
2006	26	1	0	upper middle	6195	26. 14	64. 33		22142. 67	19870. 74	12. 27	6443. 00	7. 17
2007	27	1	0	upper middle	7241	26. 63	65. 08	91	21783. 35	21983. 32	11. 40	6491. 00	6. 66
2008	28	1	0	upper middle	8487	27. 11	65. 84	91. 08	21403. 30	28363. 01	12. 33	7461. 00	7. 63
2009	29	1	0	upper middle	7312	27. 61	66. 59	91. 13	21023. 26	17141. 10	8. 49	8127. 00	7. 01
2010	30	1	0	upper middle	9069	28. 11	67. 34	91. 28	20643. 22	21205. 09	8. 32	7478. 00	5. 99
2011	31	1	0	upper middle	10428	28. 64	67. 38	90. 71	20263. 17	27062. 44	9. 06	8680. 00	6. 71
2012	32	1	0	upper middle	10835	29. 17	67. 42	90. 89	19883. 13	28023. 69	8. 87	9922. 00	7. 38
2013	33	1	0	upper middle	10974	29. 71	67. 47	90. 54	19535. 31	25586. 21	7. 85	11867. 00	8. 58
2014	34	1	0	upper middle	11307	30. 23	67. 51	90. 26	19187. 50	24868. 72	7. 28	14316. 00	9. 71
2015	35	1	0	upper middle	9766	30. 72	67. 55	90. 21	18787. 89	13320. 98	4. 44	14948. 00	9. 96
2016	36	1	0	upper middle	9508	31. 19	67. 55			12739. 67	4. 30		

Year	Renewable capacity	Renewable capacity	Energy efficiency	Green transportation	Green Building	R&D Technology Competitiveness	R&D Technology Competitiveness	HDI	Inequality	Access to electricity	CO_2	PM2.5 concentration	Air pollution-average exposure to PM2.5	Missing data 15-in total	Percentage of overall missing data
	renew_ c	renew_ c_ pct	ee	trans	bldg	rd	rd_ pct	hdi	inequality	electricity	co2	pm_ con	pm_ expo		
2005	2070. 00	9. 73	0. 46					0. 74		98. 02		10. 50		0	11. 3%
2006	2063. 00	9. 14	0. 41	1484. 60	170. 40	989. 26	0. 61	0. 75		98. 24	6159. 50	11. 80			
2007	2120. 00	9. 23	0. 38	1553. 20	223. 20	1348. 81	0. 70	0. 76	46. 05	98. 48	6659. 50	10. 80	89. 14		
2008	2120. 00	9. 08	0. 33	1603. 50	198. 20	1814. 00	0. 79	0. 77	46. 16	98. 73	6995. 90	9. 64	86. 30		
2009	2120. 00	8. 35	0. 36	1545. 10	178. 80	2038. 29	1. 01	0. 77	46. 26	99. 30	6137. 70	9. 16	88. 67		
2010	2120. 78	8. 35	0. 29	1568. 10	197. 70	2642. 89	1. 04	0. 78	45. 21	99. 29	6750. 70	9. 69	90. 10		
2011	3015. 78	10. 49	0. 26	1561. 60	177. 60	3084. 88	1. 03	0. 78	44. 15	99. 57	6700. 60	10. 10	90. 10		
2012	3371. 50	11. 55	0. 25	1537. 10	150. 90	3454. 40	1. 09	0. 79	43. 10	99. 80	6648. 50	10. 60	90. 10		
2013	4091. 96	12. 86	0. 27	1985. 70	153. 40	3833. 15	1. 18	0. 79	41. 60	99. 93	7094. 60	9. 66	89. 20		
2014	4928. 00	16. 44	0. 26	2191. 20	139. 50	4302. 13	1. 26	0. 80	40. 10	99. 99	7374. 60	9. 53	86. 13		
2015	5938. 80	19. 51	0. 20	2024. 30	131. 10	3895. 02	1. 30	0. 78	40. 06	100. 00	7266. 60	9. 39	83. 49		
2016								0. 80		100. 00					

17. Norway

Year	id	along BR	OECD	income group	GDP per capita	Population	Forest area	Biodiver-ity & Habitat	Water resources	Natural resources	Natural resources	Renewable power generation	Renewable power generation
Year	id	brc	oecd	inc_ group	gdppc	pop	forest	bio	water	natural	natural_ pct	renew_ g	renew_ g_ pct
2005	301	0	1	high	66775	4. 62	33. 11			81. 77	0. 03	137279. 96	99. 47
2006	302	0	1	high	74115	4. 66	33. 11		82040. 98	118. 58	0. 03	120750. 54	99. 32
2007	303	0	1	high	85171	4. 71	33. 12	74. 65	81118. 62	176. 22	0. 04	136004. 65	99. 13
2008	304	0	1	high	97008	4. 77	33. 12	78. 13	80118. 35	332. 86	0. 07	141285. 41	99. 40
2009	305	0	1	high	80067	4. 83	33. 13	78. 16	79118. 07	287. 05	0. 07	127253. 77	96. 57
2010	306	0	1	high	87770	4. 89	33. 13	78. 16	78117. 80	272. 48	0. 06	118363. 85	95. 73
2011	307	0	1	high	100711	4. 95	33. 14	79. 81	77117. 53	374. 51	0. 08	123157. 76	96. 49
2012	308	0	1	high	101668	5. 02	33. 14	80. 07	76117. 25	301. 40	0. 06	144695. 08	97. 95
2013	309	0	1	high	103059	5. 08	33. 15	80. 6	75238. 18	285. 90	0. 05	130902. 48	97. 70
2014	310	0	1	high	97200	5. 14	33. 16	80. 83	74359. 11	439. 53	0. 09	138639. 60	97. 66
2015	311	0	1	high	74498	5. 19	33. 16	81. 11	73285. 25	269. 27	0. 07	141699. 83	97. 71
2016	312	0	1	high	70890	5. 23				307. 03	0. 08		

Year	Renewable capacity	Renewable capacity	Energy efficiency	Green transportation	Green Building	R&D Technology Competitiveness	R&D Technology Competitiveness	HDI	Inequality	Access to electricity	CO_2	PM2. 5 concentration	Air pollution-average exposure to PM2. 5	Missing data 15-in total	Percentage of overall missing data
	renew_ c	renew_ c_ pct	ee	trans	bldg	rd	rd_ pct	hdi	inequality	electricity	co2	pm_ con	pm_ expo		
2005	28000. 00	96. 15	0. 09			4576. 43	1. 48	0. 94	32. 27	100. 00		8. 81	96. 47	0	10. 4%
2006	28152. 00	96. 06	0. 08	13. 70	318. 40	5018. 67	1. 45	0. 94	27. 29	100. 00	7606. 10		96. 59		
2007	28437. 00	94. 79	0. 07	14. 30	286. 40	6275. 70	1. 56	0. 94	28. 07	100. 00	7650. 80		97. 03		
2008	28920. 00	94. 64	0. 07	13. 90	253. 00	7182. 31	1. 55	0. 94	27. 14	100. 00	7402. 60		97. 03		
2009	29074. 00	93. 02	0. 08	13. 60	267. 50	6663. 98	1. 72	0. 94	26. 39	100. 00	7387. 20		97. 03		
2010	29059. 00	91. 70	0. 08	14. 20	285. 20	7084. 65	1. 65	0. 94	25. 86	100. 00	7671. 40	8. 94	97. 45		
2011	29422. 00	91. 71	0. 06	13. 70	235. 60	8124. 42	1. 63	0. 94	25. 54	100. 00	7320. 70	8. 66	97. 45		
2012	30155. 00	91. 77	0. 06	13. 40	207. 60	8253. 62	1. 62	0. 95	25. 90	100. 00	7083. 20	8. 57	97. 45		
2013	30792. 00	91. 95	0. 06	13. 90	207. 80	8653. 70	1. 65	0. 95	24. 05	100. 00	6904. 70	8. 69	97. 45		
2014	31040. 00	92. 12	0. 06	14. 00	168. 90	8566. 20	1. 72	0. 95	23. 33	100. 00	6890. 80	8. 87	98. 56		
2015	31180. 00	92. 15	0. 05	14. 30	148. 20	7474. 24	1. 93	0. 94	22. 61	100. 00	7074. 10	7. 91	98. 56		
2016								0. 95		100. 00		7. 87			

18. Poland

Year	id	along BR	OECD	income group	GDP per capita	Population	Forest area	Biodiver-ity & Habitat	Water resources	Natural resources	Natural resources	Renewable power generation	Renewable power generation
Year	id	brc	oecd	inc_ group	gdppc	pop	forest	bio	water	natural	natural_ pct	renew_ g	renew_ g_ pct
2005	169	1	1	high	7976	38. 17	30. 03			4172. 17	1. 37	3886. 05	2. 48
2006	170	1	1	high	9000	38. 14	30. 12		1405. 30	6184. 69	1. 80	4317. 10	2. 67
2007	171	1	1	high	11248	38. 12	30. 20	99. 7	1406. 07	6452. 43	1. 50	5450. 08	3. 42
2008	172	1	1	high	13906	38. 13	30. 29	99. 69	1406. 49	11890. 48	2. 24	6633. 41	4. 27
2009	173	1	1	high	11441	38. 15	30. 37	99. 82	1406. 91	5299. 26	1. 21	8712. 40	5. 74
2010	174	1	1	high	12598	38. 04	30. 46	99. 81	1407. 34	7638. 80	1. 59	10927. 37	6. 93
2011	175	1	1	high	13891	38. 06	30. 53	99. 81	1407. 76	10315. 71	1. 95	13171. 63	8. 05
2012	176	1	1	high	13142	38. 06	30. 60	99. 81	1408. 19	7227. 58	1. 44	16923. 67	10. 44
2013	177	1	1	high	13777	38. 04	30. 67	99. 81	1409. 14	5519. 34	1. 05	17125. 06	10. 41
2014	178	1	1	high	14337	38. 01	30. 74	99. 87	1410. 09	5189. 94	0. 95	19911. 98	12. 52
2015	179	1	1	high	12495	37. 99	30. 81	99. 87	1410. 19	3923. 49	0. 83	22766. 23	13. 80
2016	180	1	1	high	12414	37. 97	30. 81			3845. 12	0. 82		

Year	Renewable capacity	Renewable capacity	Energy efficiency	Green transportation	Green Building	R&D Tec-hnology Competitiveness	R&D Technology Competitiveness	HDI	Inequality	Access to electricity	CO_2	PM2. 5 concentration	Air pollution-average exposure to PM2. 5	Missing data 15-in total	Percentage of overall missing data
	renew_ c	renew_ c_ pct	ee	trans	bldg	rd	rd_ pct	hdi	inequality	electricity	co2	pm_ con	pm_ expo		
2005	2441. 00	7. 57	0. 30			1714. 08	0. 56	0. 81	35. 60	100. 00		19. 50		0	9. 6%
2006	2502. 00	7. 73	0. 28	1003. 00	1227. 90	1890. 44	0. 55	0. 82	33. 30	100. 00	8086. 20	19. 10			
2007	2633. 00	8. 10	0. 22	1110. 10	1131. 10	2404. 34	0. 56	0. 82	32. 20	100. 00	8042. 80	18. 80	50. 93		
2008	2860. 00	8. 75	0. 18	1161. 30	1184. 80	3167. 22	0. 60	0. 83	32. 00	100. 00	7916. 20	15. 50	51. 86		
2009	3046. 00	9. 22	0. 22	1170. 90	1214. 70	2889. 65	0. 66	0. 84	31. 40	100. 00	7643. 00	14. 90	50. 98		
2010	3449. 00	10. 34	0. 21	1222. 90	1352. 90	3442. 70	0. 72	0. 84	31. 10	100. 00	7985. 90	14. 90	56. 20		
2011	4145. 00	12. 00	0. 19	1238. 30	1197. 10	3935. 84	0. 74	0. 84	31. 10	100. 00	7871. 10	16. 70	58. 95		
2012	4914. 00	13. 92	0. 20	1190. 10	1234. 90	4404. 74	0. 88	0. 85	30. 90	100. 00	7698. 40	15. 70	60. 26		
2013	5783. 00	16. 13	0. 19	1121. 10	1189. 50	4553. 56	0. 87	0. 84	30. 70	100. 00	7587. 30	14. 16	50. 80		
2014	6199. 00	17. 21	0. 17	1136. 20	1091. 10	5122. 88	0. 94	0. 86	30. 80	100. 00	7253. 00	13. 55	59. 24		
2015	7255. 00	19. 42	0. 20	1206. 20	1080. 00	4762. 21	1. 00	0. 84	30. 60	100. 00	7343. 60	12. 95	65. 00		
2016								0. 87		100. 00					

19. Romania

Year	id	along BR	OECD	income group	GDP per capita	Population	Forest area	Biodiver-ity & Habitat	Water resources	Natural resources	Natural resources	Renewable power generation	Renewable power generation
Year	id	brc	oecd	inc_ group	gdppc	pop	forest	bio	water	natural	natural_ pct	renew_ g	renew_ g_ pct
2005	229	1	0	upper middle	4676	21. 32	27. 79			2296. 57	2. 30	20213. 00	34. 02
2006	230	1	0	upper middle	5829	21. 19	27. 90		2018. 36	2783. 88	2. 25	18360. 00	29. 28
2007	231	1	0	upper middle	8214	20. 88	28. 02	75. 33	2029. 40	2911. 40	1. 70	16004. 00	25. 95
2008	232	1	0	upper middle	10137	20. 54	28. 12	88. 52	2046. 10	4399. 91	2. 11	17224. 00	26. 52
2009	233	1	0	upper middle	8220	20. 37	28. 21	91. 87	2062. 79	2594. 12	1. 55	15626. 53	26. 94
2010	234	1	0	upper middle	8298	20. 25	28. 32	92. 01	2079. 48	2596. 00	1. 55	20420. 56	33. 49
2011	235	1	0	upper middle	9200	20. 15	28. 61	91. 71	2096. 18	4029. 15	2. 17	16372. 37	26. 31
2012	236	1	0	upper middle	8558	20. 06	28. 93	91. 64	2112. 87	4019. 22	2. 34	14994. 82	25. 40
2013	237	1	0	upper middle	9585	19. 98	29. 22	95. 53	2120. 78	3572. 62	1. 87	20269. 82	34. 42
2014	238	1	0	upper middle	10012	19. 91	29. 52	95. 53	2128. 69	3298. 68	1. 65	27325. 21	41. 61
2015	239	1	0	upper middle	8973	19. 82	29. 82	95. 71	2150. 71	1840. 32	1. 04	26350. 65	39. 75
2016	240	1	0	upper middle	9523	19. 70	29. 82			1722. 64	0. 92		

Year	Renewable capacity	Renewable capacity	Energy efficiency	Green transpor-tation	Green Building	R&D Tec-hnology Competi-tiveness	R&D Technology Competi-tiveness	HDI	Inequality	Access to electricity	CO_2	PM2.5 concen-tration	Air pollution-average exposure to PM2.5	Missing data 15-in total	Percentage of overall missing data
	renew_ c	renew_ c_ pct	ee	trans	bldg	rd	rd_ pct	hdi	inequality	electricity	co2	pm_ con	pm_ expo		
2005	6116.00	31.90	0.39			406.92	0.41	0.77	31.00	100.00		16.70		0	9.6%
2006	6285.00	32.48	0.32	590.60	541.00	557.43	0.45	0.78	40.20	100.00	4497.40	17.70			
2007	6291.00	31.84	0.23	637.00	462.70	893.52	0.52	0.80	38.30	100.00	4428.20	17.50	59.07		
2008	6307.00	32.10	0.19	728.80	402.70	1184.22	0.57	0.80	35.90	100.00	4443.10	14.80	56.87		
2009	6381.00	32.64	0.21	729.00	406.50	772.20	0.46	0.80	34.50	100.00	3803.50	13.90	55.72		
2010	6773.00	34.02	0.21	682.40	406.60	759.29	0.45	0.80	33.50	100.00	3692.30	14.30	62.10		
2011	7375.00	35.98	0.19	697.10	409.40	914.29	0.49	0.80	33.50	100.00	4015.30	15.40	63.41		
2012	7898.00	36.27	0.20	764.50	429.60	828.81	0.48	0.80	34.00	100.00	3919.30	14.90	61.49		
2013	9499.00	41.24	0.17	747.30	422.40	741.18	0.39	0.80	34.60	100.00	3452.60	13.80	58.65		
2014	10504.00	43.67	0.16	768.50	403.70	762.58	0.38	0.81	35.00	100.00	3425.50	13.38	64.22		
2015	10629.00	44.39	0.12	779.20	417.50	867.05	0.49	0.80	37.40	100.00	3508.80	12.97	69.34		
2016								0.81		100.00					

20. Russia

Year	id	along BR	OECD	income group	GDP per capita	Population	Forest area	Biodiver-ity & Habitat	Water resources	Natural resources	Natural resources	Renewable power generation	Renewable power generation
Year	id	brc	oecd	inc_ group	gdppc	pop	forest	bio	water	natural	natural_ pct	renew_ g	renew_ g_ pct
2005	157	1	0	upper middle	5324	143. 52	49. 37					173485. 76	18. 20
2006	158	1	0	upper middle	6920	143. 05	49. 46		30242. 56			174202. 46	17. 49
2007	159	1	0	upper middle	9101	142. 81	49. 54	74. 84	30195. 00			177892. 85	17. 52
2008	160	1	0	upper middle	11635	142. 74	49. 62	74. 15	30178. 28			165567. 01	15. 91
2009	161	1	0	upper middle	8563	142. 79	49. 70	73. 89	30161. 55			175025. 41	17. 64
2010	162	1	0	upper middle	10675	142. 85	49. 77	73. 41	30144. 83			167336. 55	16. 12
2011	163	1	0	upper middle	14212	142. 96	49. 77	73. 17	30128. 10			166684. 77	15. 80
2012	164	1	0	upper middle	15155	143. 20	49. 77	73. 15	30111. 38			166646. 16	15. 56
2013	165	1	0	upper middle	15544	143. 51	49. 77	72. 93	30046. 69			181895. 13	17. 17
2014	166	1	0	upper middle	13902	143. 82	49. 76	73. 36	29981. 99			176320. 49	16. 57
2015	167	1	0	upper middle	9057	144. 10	49. 76	72. 99	29994. 39			169267. 59	15. 86
2016	168	1	0	upper middle	8748	144. 34	49. 76						

Year	Renewable capacity	Renewable capacity	Energy efficiency	Green transpor-tation	Green Building	R&D Tec-hnology Competi-tiveness	R&D Technology Competi-tiveness	HDI	Inequality	Access to electricity	CO_2	PM2. 5 concen-tration	Air pollution-average exposure to PM2. 5	Missing data 15-in total	Percentage of overall missing data
	renew_ c	renew_ c_ pct	ee	trans	bldg	rd	rd_ pct	hdi	inequality	electricity	co2	pm_ con	pm_ expo		
2005	45391. 00	20. 72						0. 76	44. 50	100. 00		10. 60		3	33. 3%
2006	45572. 00	20. 58		1601. 70	1014. 60			0. 77	45. 10	100. 00	10747. 60	12. 20			
2007	46311. 00	20. 68		1615. 30	1036. 50			0. 77	35. 00	100. 00	10738. 00	12. 10	88. 46		
2008	46521. 00	20. 65		1718. 20	1091. 10			0. 77	41. 42	100. 00	10883. 70	9. 99	83. 48		
2009	46759. 00	20. 74		1598. 80	979. 60			0. 78	39. 69	100. 00	10087. 80	9. 49	84. 64		
2010	46881. 00	21. 03		1708. 70	959. 00			0. 79	34. 20	100. 00	10703. 00	9. 94	88. 91		
2011	46930. 00	21. 05		1746. 20	973. 60			0. 80	41. 04	100. 00	11222. 70	10. 10	88. 91		
2012	49151. 00	21. 12		1656. 90	855. 10			0. 80	41. 59	100. 00	10829. 60	9. 84	88. 91		
2013	49662. 00	20. 74		1663. 10	852. 40			0. 81	33. 70	100. 00	10547. 00	9. 31	88. 91		
2014	50686. 00	19. 57		1678. 10	965. 30			0. 81	35. 51	100. 00	10338. 40	9. 04	89. 67		
2015	51068. 00	19. 87		1669. 50	1032. 10			0. 79	34. 70	100. 00	10194. 40	8. 77	89. 67		
2016								0. 82		100. 00					

21. Saudi Arabia

Year	id	along BR	OECD	income group	GDP per capita	Population	Forest area	Biodiver-ity & Habitat	Water resources	Natural resources	Natural resources	Renewable power generation	Renewable power generation
Year	id	brc	oecd	inc_ group	gdppc	pop	forest	bio	water	natural	natural_ pct	renew_ g	renew_ g_ pct
2005	85	1	0	high	13274	23. 91	0. 45			159356. 28	50. 22		
2006	86	1	0	high	14827	24. 58	0. 45		97. 39	183729. 93	50. 42		
2007	87	1	0	high	15947	25. 25	0. 45	83. 12	95. 04	188340. 79	46. 77		
2008	88	1	0	high	19437	25. 94	0. 45	83. 06	92. 53	272081. 61	53. 96		
2009	89	1	0	high	15655	26. 66	0. 45	82. 56	90. 03	141187. 84	33. 83		
2010	90	1	0	high	18754	27. 43	0. 45	81. 99	87. 52	210541. 27	40. 93		
2011	91	1	0	high	23256	28. 24	0. 45	81. 61	85. 02	328886. 08	50. 08		
2012	92	1	0	high	24883	29. 09	0. 45	81. 53	82. 51	347802. 49	48. 05		1. 00
2013	93	1	0	high	24646	29. 94	0. 45	81. 37	80. 25	329941. 89	44. 71		1. 00
2014	94	1	0	high	24407	30. 78	0. 45	81. 29	77. 98	301406. 96	40. 13		1. 00
2015	95	1	0	high	20482	31. 56	0. 45	80. 75	75. 33	152125. 92	23. 54		
2016	96	1	0	high	20029	32. 28	0. 45			175822. 69	27. 20		

Year	Renewable capacity	Renewable capacity	Energy efficiency	Green transportation	Green Building	R&D Tec-hnology Competitiveness	R&D Technology Competitiveness	HDI	Inequality	Access to electricity	CO_2	PM2. 5 concentration	Air pollution-average exposure to PM2. 5	Missing data 15-in total	Percentage of overall missing data
	renew_ c	renew_ c_ pct	ee	trans	bldg	rd	rd_ pct	hdi	inequality	electricity	co2	pm_ con	pm_ expo		
2005			0. 39			134. 32	0. 04	0. 78		100. 00		10. 10	57. 75	1	29. 2%
2006				0. 37	3219. 60	155. 20	154. 84	0. 04	0. 78		100. 00	12453. 50	10. 10	56. 83	
2007				0. 35	3497. 90	158. 40	181. 95	0. 05	0. 79		100. 00	12770. 20	10. 60	55. 44	
2008				0. 31	3683. 60	162. 50	247. 06	0. 05	0. 80		100. 00	13617. 10	11. 30	55. 44	
2009			0. 40	3704. 30	159. 80	306. 11	0. 07	0. 81		100. 00	13842. 30	11. 60	55. 44		
2010			0. 36	3751. 30	157. 00	4548. 52	0. 88	0. 82		100. 00	14919. 30	11. 80	51. 82		
2011			0. 27	3845. 90	178. 10	5912. 98	0. 90	0. 84		100. 00	15095. 30	11. 60	51. 81		
2012	0. 00	0. 28	4083. 70	156. 80	6369. 10	0. 88	0. 84		100. 00	15708. 50	11. 90	51. 36			
2013	0. 00	0. 26	4118. 50	141. 70	6037. 38	0. 82	0. 85		100. 00	15597. 00	12. 41	51. 42			
2014	0. 00	0. 28	4252. 30	134. 50	8226. 00	1. 10	0. 85		100. 00	16401. 50	12. 70	48. 61			
2015			0. 25	4506. 30	146. 50	7957. 39	1. 23	0. 82		100. 00	16850. 30	12. 98	48. 96		
2016								0. 85		100. 00					

22. Singapore

Year	id	along BR	OECD	income group	GDP per capita	Population	Forest area	Biodiver-ity & Habitat	Water resources	Natural resources	Natural resources	Renewable power generation	Renewable power generation
Year	id	brc	oecd	inc_ group	gdppc	pop	forest	bio	water	natural	natural_ pct	renew_ g	renew_ g_ pct
2005	13	1	0	high	29870	4. 27	23. 73			0. 40	0. 00	478. 00	1. 25
2006	14	1	0	high	33580	4. 40	23. 56		133. 08	0. 48	0. 00	477. 00	1. 21
2007	15	1	0	high	39224	4. 59	23. 53	74. 21	130. 76	0. 72	0. 00	487. 00	1. 18
2008	16	1	0	high	39721	4. 84	23. 36	73. 87	127. 20	1. 20	0. 00	525. 00	1. 26
2009	17	1	0	high	38578	4. 99	23. 36	73. 55	123. 63	1. 12	0. 00	535. 00	1. 30
2010	18	1	0	high	46570	5. 08	23. 29	73. 13	120. 07	1. 09	0. 00	592. 00	1. 31
2011	19	1	0	high	53094	5. 18	23. 22	72. 6	116. 51	1. 43	0. 00	611. 00	1. 33
2012	20	1	0	high	54451	5. 31	23. 16	72. 28	112. 94	1. 16	0. 00	638. 00	1. 36
2013	21	1	0	high	55618	5. 40	23. 13	71. 81	111. 32	1. 02	0. 00	742. 00	1. 55
2014	22	1	0	high	56007	5. 47	23. 06	72	109. 69	1. 53	0. 00	821. 00	1. 66
2015	23	1	0	high	52889	5. 54	23. 06	71. 48	104. 95	1. 10	0. 00	918. 00	1. 82
2016	24	1	0	high	52962	5. 61	23. 06			1. 38	0. 00		

Year	Renewable capacity	Renewable capacity	Energy efficiency	Green transpor-tation	Green Building	R&D Tec-hnology Competi-tiveness	R&D Technology Competi-tiveness	HDI	Inequality	Access to electricity	CO_2	PM2. 5 concen-tration	Air pollution-average exposure to PM2. 5	Missing data 15-in total	Percentage of overall missing data
	renew_ c	renew_ c_ pct	ee	trans	bldg	rd	rd_ pct	hdi	inequality	electricity	co2	pm_ con	pm_ expo		
2005			0. 17			2754. 04	2. 16	0. 87		100. 00		5. 22		0	16. 3%
2006			0. 16	1320. 40	104. 90	3150. 73	2. 13	0. 88		100. 00	8698. 60	5. 86			
2007			0. 12	1308. 70	124. 40	4210. 03	2. 34	0. 88		100. 00	8589. 30	5. 65	92. 74		
2008			0. 13	1265. 80	122. 80	5037. 50	2. 62	0. 88		100. 00	8132. 30	5. 20	92. 74		
2009			0. 11	1343. 70	124. 70	4152. 47	2. 16	0. 91	47. 10	100. 00	8086. 20	4. 26	92. 74		
2010			0. 11	1384. 90	125. 70	4764. 44	2. 02	0. 91	47. 20	100. 00	8721. 00	4. 51	89. 25		
2011	4. 60	0. 05	0. 10	1456. 30	120. 30	5924. 82	2. 15	0. 92	47. 30	100. 00	9015. 40	4. 55	89. 25		
2012	7. 80	0. 08	0. 09	1384. 50	117. 70	5805. 19	2. 01	0. 92	47. 40	100. 00	8651. 30	5. 16	89. 25		
2013	11. 80	0. 10	0. 09	1316. 80	120. 00	6042. 37	2. 01	0. 93	47. 50	100. 00	8565. 10	4. 44	89. 25		
2014	25. 50	0. 20	0. 09	1257. 30	110. 10	6732. 04	2. 20	0. 93	47. 60	100. 00	8286. 10	4. 31	82. 33		
2015	45. 80	0. 35	0. 07	1199. 40	110. 10	6017. 44	2. 06	0. 91	47. 70	100. 00	8027. 50	4. 17	82. 33		
2016								0. 93		100. 00					

23. Slovakia

Year	id	along BR	OECD	income group	GDP per capita	Population	Forest area	Biodiver-ity & Habitat	Water resources	Natural resources	Natural resources	Renewable power generation	Renewable power generation
Year	id	brc	oecd	inc_ group	gdppc	pop	forest	bio	water	natural	natural_ pct	renew_ g	renew_ g_ pct
2005	205	1	1	high	11631	5. 37	40. 17					4691. 36	14. 91
2006	206	1	1	high	13100	5. 37	40. 20		2346. 90			4828. 67	15. 37
2007	207	1	1	high	16015	5. 37	40. 22	100	2344. 35			4963. 01	17. 69
2008	208	1	1	high	18604	5. 38	40. 25	100	2341. 49			4595. 05	15. 87
2009	209	1	1	high	16460	5. 39	40. 29	100	2338. 64			4955. 72	18. 95
2010	210	1	1	high	16555	5. 39	40. 32	100	2335. 78			6025. 22	21. 63
2011	211	1	1	high	18139	5. 40	40. 33	100	2332. 92			5064. 03	17. 67
2012	212	1	1	high	17207	5. 41	40. 33	100	2330. 06			5538. 93	19. 32
2013	213	1	1	high	18109	5. 41	40. 33	100	2327. 68			6423. 85	22. 28
2014	214	1	1	high	18501	5. 42	40. 34	100	2325. 30			6287. 05	22. 94
2015	215	1	1	high	15963	5. 42	40. 35	100	2322. 16			6101. 46	22. 68
2016	216	1	1	high	16530	5. 43	40. 35						

Year	Renewable capacity	Renewable capacity	Energy efficiency	Green transportation	Green Building	R&D Technology Competitiveness	R&D Technology Competitiveness	HDI	Inequality	Access to electricity	CO_2	PM2. 5 concentration	Air pollution-average exposure to PM2. 5	Missing data 15-in total	Percentage of overall missing data
	renew_ c	renew_ c_ pct	ee	trans	bldg	rd	rd_ pct	hdi	inequality	electricity	co2	pm_ con	pm_ expo		
2005	2488. 00	30. 13						0. 80	26. 20	100. 00		17. 40		3	33. 3%
2006	2488. 00	30. 27		1133. 40	1120. 10			0. 81	28. 10	100. 00	6765. 60	17. 70			
2007	2490. 00	33. 89		1239. 70	1042. 40			0. 82	24. 50	100. 00	6613. 80	17. 50	59. 35		
2008	2523. 00	34. 29		1355. 10	1140. 90			0. 82	23. 70	100. 00	6551. 10	14. 60	59. 59		
2009	2462. 00	34. 42		1166. 60	1201. 10			0. 83	24. 80	100. 00	6017. 30	14. 10	59. 58		
2010	2493. 00	31. 67		1317. 10	1208. 00			0. 84	25. 90	100. 00	6365. 00	14. 10	66. 46		
2011	2602. 00	31. 11		1318. 40	951. 20			0. 84	25. 70	100. 00	6084. 80	15. 30	65. 81		
2012	2598. 00	30. 88		1196. 10	869. 60			0. 84	25. 30	100. 00	5779. 10	15. 10	65. 83		
2013	2598. 00	30. 72		1195. 90	986. 80			0. 85	24. 20	100. 00	5870. 20	13. 65	63. 06		
2014	2596. 00	32. 08		1117. 60	783. 10			0. 85	26. 10	100. 00	5381. 20	13. 18	64. 08		
2015	2599. 00	33. 39		1100. 10	798. 50			0. 84	23. 70	100. 00	5428. 80	12. 72	66. 57		
2016								0. 86		100. 00					

24. Slovenia

Year	id	along BR	OECD	income group	GDP per capita	Population	Forest area	Biodiver-ity & Habitat	Water resources	Natural resources	Natural resources	Renewable power generation	Renewable power generation
Year	id	brc	oecd	inc_ group	gdppc	pop	forest	bio	water	natural	natural_ pct	renew_ g	renew_ g_ pct
2005	217	1	1	high	18169	2. 00	61. 72			68. 62	0. 19	3575. 00	23. 65
2006	218	1	1	high	19726	2. 01	61. 76		9270. 53	83. 71	0. 21	3703. 00	24. 50
2007	219	1	1	high	23841	2. 02	61. 80	100	9251. 18	109. 36	0. 23	3379. 00	22. 46
2008	220	1	1	high	27502	2. 02	61. 84	100	9216. 06	216. 02	0. 39	4308. 00	26. 27
2009	221	1	1	high	24634	2. 04	61. 88	100	9180. 95	99. 00	0. 20	4907. 00	29. 92
2010	222	1	1	high	23439	2. 05	61. 92	100	9145. 84	133. 08	0. 28	4803. 05	29. 92
2011	223	1	1	high	24984	2. 05	61. 93	100	9110. 73	181. 27	0. 35	3915. 11	24. 38
2012	224	1	1	high	22478	2. 06	61. 94	100	9075. 62	112. 43	0. 24	4381. 34	27. 84
2013	225	1	1	high	23144	2. 06	61. 95	100	9065. 01	93. 48	0. 20	5205. 03	32. 32
2014	226	1	1	high	24002	2. 06	61. 96	100	9054. 40	129. 02	0. 26	6716. 54	38. 52
2015	227	1	1	high	20713	2. 06	61. 97	100	9004. 42	106. 72	0. 25	4438. 18	29. 39
2016	228	1	1	high	21650	2. 07	61. 97			114. 37	0. 26		

Year	Renewable capacity	Renewable capacity	Energy efficiency	Green transportation	Green Building	R&D Technology Competitiveness	R&D Technology Competitiveness	HDI	Inequality	Access to electricity	CO_2	PM2. 5 concentration	Air pollution-average exposure to PM2. 5	Missing data 15-in total	Percentage of overall missing data
	renew_ c	renew_ c_ pct	ee	trans	bldg	rd	rd_ pct	hdi	inequality	electricity	co2	pm_ con	pm_ expo		
2005	875. 00	29. 24	0. 20			513. 33	1. 41	0. 87	23. 80	100. 00		17. 20		0	9. 6%
2006	905. 00	29. 79	0. 18	2265. 20	977. 50	606. 84	1. 53	0. 87	23. 70	100. 00	7871. 80	18. 10			
2007	906. 00	29. 85	0. 15	2550. 70	722. 40	685. 07	1. 42	0. 88	23. 20	100. 00	7805. 90	17. 80	61. 58		
2008	914. 00	30. 58	0. 14	2994. 80	910. 70	903. 69	1. 63	0. 88	23. 40	100. 00	8311. 00	12. 40	59. 34		
2009	955. 00	31. 31	0. 14	2486. 10	866. 90	912. 59	1. 82	0. 88	22. 70	100. 00	7402. 60	12. 40	58. 28		
2010	1139. 00	35. 67	0. 15	2542. 80	907. 40	988. 00	2. 06	0. 88	23. 80	100. 00	7542. 10	12. 90	75. 26		
2011	1139. 00	34. 85	0. 14	2710. 80	777. 60	1243. 00	2. 42	0. 88	23. 80	100. 00	7517. 30	14. 40	76. 36		
2012	1143. 00	34. 11	0. 15	2713. 40	638. 20	1192. 28	2. 58	0. 89	23. 70	100. 00	7226. 80	13. 70	74. 48		
2013	1193. 00	34. 74	0. 14	2566. 00	580. 90	1241. 11	2. 60	0. 89	24. 40	100. 00	6902. 10	11. 77	69. 11		
2014	1190. 00	34. 45	0. 13	2571. 50	465. 30	1180. 18	2. 38	0. 89	25. 00	100. 00	6188. 10	11. 08	75. 45		
2015	1191. 00	35. 45	0. 15	2556. 40	509. 20	945. 34	2. 21	0. 88	24. 50	100. 00	6221. 80	10. 40	75. 82		
2016								0. 90		100. 00					

25. Switzerland

Year	id	along BR	OECD	income group	GDP per capita	Population	Forest area	Biodiver-ity & Habitat	Water resources	Natural resources	Natural resources	Renewable power generation	Renewable power generation
Year	id	brc	oecd	inc_ group	gdppc	pop	forest	bio	water	natural	natural_ pct	renew_ g	renew_ g_ pct
2005	313	0	1	high	54953	7. 44	30. 80			19571. 95	4. 79	33318. 97	55. 86
2006	314	0	1	high	57580	7. 48	30. 89		5409. 54	24318. 84	5. 64	33136. 22	51. 73
2007	315	0	1	high	63555	7. 55	30. 98	84. 91	5350. 20	34392. 66	7. 17	37295. 45	54. 91
2008	316	0	1	high	72488	7. 65	31. 07	84. 57	5290. 56	72029. 58	12. 99	38382. 34	55. 68
2009	317	0	1	high	69927	7. 74	31. 16	84. 65	5230. 91	32036. 75	5. 92	38023. 91	55. 54
2010	318	0	1	high	74606	7. 82	31. 25	84. 56	5171. 27	44769. 98	7. 67	38471. 71	56. 73
2011	319	0	1	high	88416	7. 91	31. 35	86. 91	5111. 63	60422. 00	8. 64	34960. 84	54. 09
2012	320	0	1	high	83538	8. 00	31. 45	89. 75	5051. 98	47852. 04	7. 16	41540. 04	59. 47
2013	321	0	1	high	85112	8. 09	31. 54	89. 57	4992. 82	47404. 05	6. 89	41570. 05	59. 19
2014	322	0	1	high	86606	8. 19	31. 64	89. 95	4933. 66	40831. 32	5. 76	41635. 96	58. 02
2015	323	0	1	high	82016	8. 28	31. 73	89. 99	4873. 72	29337. 70	4. 32	42120. 47	62. 20
2016	324	0	1	high	79866	8. 37				31419. 03	4. 70		

Year	Renewable capacity	Renewable capacity	Energy efficiency	Green transportation	Green Building	R&D Technology Competitiveness	R&D Technology Competitiveness	HDI	Inequality	Access to electricity	CO_2	PM2. 5 concentration	Air pollution-average exposure to PM2. 5	Missing data 15-in total	Percentage of overall missing data
	renew_ c	renew_ c_ pct	ee	trans	bldg	rd	rd_ pct	hdi	inequality	electricity	co2	pm_ con	pm_ expo		
2005	12844. 00	73. 65	0. 06					0. 91		100. 00		10. 91		0	19. 6%
2006	12831. 00	73. 50	0. 06	16. 70	2136. 90			0. 92		100. 00	5796. 20				
2007	12905. 00	73. 45	0. 05	17. 00	1881. 00			0. 92	34. 47	100. 00	5488. 00		71. 59		
2008	12866. 00	73. 05	0. 05	17. 40	1972. 00	15135. 90	2. 73	0. 93	33. 96	100. 00	5612. 00		67. 16		
2009	13009. 00	73. 19	0. 05	17. 20	1913. 80			0. 93	32. 96	100. 00	5377. 90		71. 65		
2010	13284. 00	73. 45	0. 04	17. 20	2039. 60			0. 93	32. 72	100. 00	5504. 70	11. 00	81. 03		
2011	13326. 00	73. 08	0. 04	17. 00	1621. 10			0. 94	31. 83	100. 00	4957. 30	11. 09	83. 17		
2012	13269. 00	71. 40	0. 04	17. 10	1763. 50	19816. 51	2. 97	0. 94	31. 64	100. 00	5083. 60	11. 17	80. 68		
2013	13292. 00	70. 21	0. 04	17. 10	1846. 90			0. 94	30. 85	100. 00	5149. 30	12. 04	78. 36		
2014	13242. 00	69. 09	0. 04	16. 90	1428. 80			0. 94	30. 26	100. 00	4623. 00	12. 33	71. 50		
2015	13364. 00	68. 12	0. 03	16. 20	1513. 90			0. 93	29. 66	100. 00	4505. 20	10. 60	68. 60		
2016								0. 94		100. 00		10. 51			

26. The United States

Year	id	along BR	OECD	income group	GDP per capita	Population	Forest area	Biodiver-ity & Habitat	Water resources	Natural resources	Natural resources	Renewable power generation	Renewable power generation
Year	id	brc	oecd	inc_ group	gdppc	pop	forest	bio	water	natural	natural_ pct	renew_ g	renew_ g_ pct
2005	325	0	1	high	44308	295. 52	33. 26			627052. 67	4. 79	368390. 88	8. 58
2006	326	0	1	high	46437	298. 38	33. 35		9424. 01	781950. 54	5. 64	397595. 96	9. 24
2007	327	0	1	high	48062	301. 23	33. 44	80. 26	9354. 94	1037530. 29	7. 17	363984. 43	8. 37
2008	328	0	1	high	48401	304. 09	33. 58	79. 64	9278. 90	1912415. 46	12. 99	393298. 19	9. 00
2009	329	0	1	high	47002	306. 77	33. 66	79. 55	9202. 85	853045. 30	5. 92	431044. 60	10. 29
2010	330	0	1	high	48375	309. 34	33. 75	79. 84	9126. 80	1147609. 04	7. 67	443112. 67	10. 12
2011	331	0	1	high	49794	311. 64	33. 78	79. 67	9050. 76	1340267. 97	8. 64	531752. 89	12. 23
2012	332	0	1	high	51451	313. 99	33. 81	79. 3	8974. 71	1157202. 67	7. 16	515196. 58	12. 01
2013	333	0	1	high	52782	316. 23	33. 84	79	8909. 52	1149223. 88	6. 89	544235. 68	12. 64
2014	334	0	1	high	54697	318. 62	33. 87	79. 23	8844. 32	1003397. 91	5. 76	562093. 76	12. 95
2015	335	0	1	high	56444	321. 04	33. 90	78. 96	8761. 69	782612. 25	4. 32	571099. 40	13. 23
2016	336	0	1	high	57589	323. 41				875016. 30	4. 70		

Year	Renewable capacity	Renewable capacity	Energy efficiency	Green transportation	Green Building	R&D Technology Competitiveness	R&D Technology Competitiveness	HDI	Inequality	Access to electricity	CO_2	PM2. 5 concentration	Air pollution-average exposure to PM2. 5	Missing data 15-in total	Percentage of overall missing data
	renew_ c	renew_ c_ pct	ee	trans	bldg	rd	rd_ pct	hdi	inequality	electricity	co2	pm_ con	pm_ expo		
2005	109588. 00	11. 20	0. 18			327970. 34	2. 50	0. 90		100. 00		10. 43	88. 08	0	9. 6%
2006	112579. 00	11. 41	0. 17	1805. 80	1731. 40	352206. 97	2. 54	0. 91		100. 00	18746. 10	10. 06	88. 65		
2007	118649. 00	11. 91	0. 16	1806. 20	1819. 20	379729. 55	2. 62	0. 91	41. 75	100. 00	18846. 30	9. 70	88. 21		
2008	126848. 00	12. 54	0. 15	1708. 10	1826. 00	407753. 29	2. 77	0. 91	41. 32	100. 00	18098. 00	9. 34	90. 27		
2009	137615. 00	13. 40	0. 15	1622. 40	1794. 00	406934. 30	2. 82	0. 91	40. 89	100. 00	16664. 10	8. 98	90. 27		
2010	143054. 00	13. 74	0. 15	1675. 40	1758. 80	409118. 45	2. 73	0. 92	40. 46	100. 00	17259. 30	8. 62	91. 58		
2011	149536. 00	14. 17	0. 14	1654. 30	1707. 70	430288. 81	2. 77	0. 92	47. 60	100. 00	16693. 10	8. 65	91. 58		
2012	165133. 00	15. 46	0. 13	1691. 80	1553. 70	435873. 63	2. 70	0. 92	47. 70	100. 00	15998. 10	8. 68	91. 58		
2013	169863. 00	15. 95	0. 13	1688. 30	1720. 30	457379. 28	2. 74	0. 92	47. 70	100. 00	16108. 70	8. 53	91. 58		
2014	178619. 00	16. 64	0. 13	1721. 30	1785. 50	480118. 43	2. 75	0. 92	49. 25	100. 00	16189. 20	8. 50	93. 23		
2015	190330. 00	17. 75	0. 12	1752. 00	1634. 80	506265. 57	2. 79	0. 92	50. 58	100. 00	15534. 50	9. 17	93. 23		
2016								0. 92		100. 00		9. 20			

27. Turkey

Year	id	along BR	OECD	income group	GDP per capita	Population	Forest area	Biodiver-ity & Habitat	Water resources	Natural resources	Natural resources	Renewable power generation	Renewable power generation
Year	id	brc	oecd	inc_ group	gdppc	pop	forest	bio	water	natural	natural_ pct	renew_ g	renew_ g_ pct
2005	73	1	1	upper middle	7117	67. 90	13. 85			1368. 90	0. 28	39748. 00	24. 54
2006	74	1	1	upper middle	7727	68. 76	13. 99		3308. 41	1999. 30	0. 38	44522. 00	25. 25
2007	75	1	1	upper middle	9310	69. 60	14. 13	28. 21	3261. 62	2753. 15	0. 42	36457. 00	19. 03
2008	76	1	1	upper middle	10382	70. 44	14. 28	27. 08	3218. 12	4683. 34	0. 64	34421. 00	17. 35
2009	77	1	1	upper middle	8624	71. 34	14. 42	26. 52	3174. 62	2358. 84	0. 38	38141. 00	19. 58
2010	78	1	1	upper middle	10112	72. 33	14. 56	23. 79	3131. 12	3937. 23	0. 54	55712. 00	26. 38
2011	79	1	1	upper middle	10538	73. 41	14. 69	23. 14	3087. 62	5464. 63	0. 71	58098. 00	25. 33
2012	80	1	1	upper middle	10539	74. 57	14. 82	22. 83	3044. 13	4482. 14	0. 57	65216. 00	27. 23
2013	81	1	1	upper middle	10800	75. 79	14. 96	22. 87	2995. 50	4000. 54	0. 49	69220. 00	28. 82
2014	82	1	1	upper middle	10304	77. 03	15. 09	22. 83	2946. 88	3363. 27	0. 42	52628. 00	20. 89
2015	83	1	1	upper middle	9130	78. 27	15. 22	19. 98	2906. 49	2115. 23	0. 30	83657. 00	31. 96
2016	84	1	1	upper middle	10863	79. 51	15. 22			2826. 09	0. 33		

Year	Renewable capacity	Renewable capacity	Energy efficiency	Green transportation	Green Building	R&D Technology Competitiveness	R&D Technology Competitiveness	HDI	Inequality	Access to electricity	CO_2	PM2. 5 concentration	Air pollution-average exposure to PM2. 5	Missing data 15-in total	Percentage of overall missing data
	renew_ c	renew_ c_ pct	ee	trans	bldg	rd	rd_ pct	hdi	inequality	electricity	co2	pm_ con	pm_ expo		
2005	12376. 00	31. 86	0. 17			2856. 10	0. 59	0. 70	42. 61	97. 22		11. 90		0	9. 6%
2006	12581. 00	31. 01	0. 18	600. 00	528. 90	3084. 31	0. 58	0. 71	44. 90	97. 73	3457. 20	12. 30			
2007	13005. 00	31. 85	0. 15	675. 50	564. 50	4681. 58	0. 72	0. 71	43. 20	98. 27	3786. 10	12. 30	78. 23		
2008	13668. 00	32. 68	0. 13	628. 10	725. 00	5300. 82	0. 72	0. 72	43. 00	98. 81	3731. 90	11. 80	75. 99		
2009	14877. 00	33. 24	0. 16	614. 10	653. 30	5220. 96	0. 85	0. 73	44. 20	99. 31	3568. 00	11. 70	77. 38		
2010	16700. 00	33. 72	0. 15	594. 10	633. 50	6161. 56	0. 84	0. 75	43. 50	100. 00	3641. 10	11. 60	73. 01		
2011	18435. 00	34. 84	0. 15	594. 80	676. 80	6637. 72	0. 86	0. 76	43. 30	99. 91	3854. 70	12. 40	73. 45		
2012	21487. 00	37. 66	0. 15	681. 40	812. 70	7229. 54	0. 92	0. 77	42. 80	99. 99	4043. 40	12. 10	75. 08		
2013	24810. 00	38. 76	0. 14	754. 60	640. 30	7719. 92	0. 94	0. 78	42. 10	100. 00	3750. 30	12. 00	71. 61		
2014	27718. 00	38. 03	0. 15	795. 00	607. 60	7981. 69	1. 01	0. 78	42. 23	100. 00	4001. 50	11. 99	75. 50		
2015	31244. 00	42. 71	0. 18	935. 80	686. 40	7565. 49	1. 06	0. 75	41. 97	100. 00	4095. 70	11. 99	76. 50		
2016								0. 79		100. 00					

28. United Arab Emirates

Year	id	along BR	OECD	income group	GDP per capita	Population	Forest area	Biodiver-ity & Habitat	Water resources	Natural resources	Natural resources	Renewable power generation	Renewable power generation
Year	id	brc	oecd	inc_ group	gdppc	pop	forest	bio	water	natural	natural_ pct	renew_ g	renew_ g_ pct
2005	97	1	0	high	40299	4. 58	3. 73			45676. 13	24. 75		
2006	98	1	0	high	42950	5. 24	3. 74		25. 49	56649. 25	25. 16		
2007	99	1	0	high	42914	6. 04	3. 76	89. 91	24. 82	59660. 41	23. 00		
2008	100	1	0	high	45720	6. 89	3. 77	89. 59	23. 22	82750. 55	26. 25		
2009	101	1	0	high	32905	7. 67	3. 78	89. 89	21. 63	43579. 85	17. 28		
2010	102	1	0	high	34342	8. 27	3. 80	89. 71	20. 04	58455. 88	20. 58		
2011	103	1	0	high	39901	8. 67	3. 81	89. 36	18. 45	95662. 47	27. 64		
2012	104	1	0	high	41712	8. 90	3. 82	90. 06	16. 85	102138. 91	27. 51		
2013	105	1	0	high	42831	9. 01	3. 83	89. 74	16. 69	98221. 69	25. 46	100. 00	0. 09
2014	106	1	0	high	43963	9. 07	3. 85	89. 68	16. 54	91107. 03	22. 85	301. 00	0. 26
2015	107	1	0	high	40438	9. 15	3. 86	89. 65	14. 07	48516. 36	13. 11	296. 00	0. 23
2016	108	1	0	high	37622	9. 27	3. 86			53245. 87	15. 27		

Year	Renewable capacity	Renewable capacity	Energy efficiency	Green transportation	Green Building	R&D Technology Competitiveness	R&D Technology Competitiveness	HDI	Inequality	Access to electricity	CO_2	PM2.5 concentration	Air pollution-average exposure to PM2.5	Missing data 15-in total	Percentage of overall missing data
	renew_ c	renew_ c_ pct	ee	trans	bldg	rd	rd_ pct	hdi	inequality	electricity	co2	pm_ con	pm_ expo		
2005			0. 24					0. 83		100. 00		10. 40		1	30. 8%
2006			0. 21	4625. 40	25. 40			0. 84		100. 00	22242. 60	10. 90			
2007			0. 19	4089. 90	28. 30			0. 84		100. 00	20657. 50	11. 10	66. 53		
2008			0. 19	3790. 70	43. 20			0. 84		100. 00	21228. 90	11. 70	66. 07		
2009			0. 24	3736. 40	63. 10			0. 84		100. 00	19270. 00	11. 90	64. 85		
2010			0. 22	3607. 80	65. 50			0. 84		100. 00	18543. 90	11. 90	59. 63		
2011	10. 00	0. 04	0. 19	3443. 50	77. 20	1692. 80	0. 49	0. 85		100. 00	18091. 20	11. 50	58. 42		
2012	10. 00	0. 04	0. 18	3435. 20	61. 30	2069. 41	0. 56	0. 85		100. 00	18806. 30	11. 70	59. 24		
2013	60. 00	0. 22	0. 18	3834. 70	96. 70	2413. 33	0. 63	0. 86		100. 00	19538. 10	12. 18	60. 41		
2014	60. 00	0. 21	0. 18	3915. 00	96. 90	2766. 89	0. 69	0. 86		100. 00	19402. 80	12. 35	59. 99		
2015	60. 00	0. 21	0. 17	3372. 00	105. 90	3204. 29	0. 87	0. 84		100. 00	19679. 60	12. 53	58. 44		
2016										0. 86		100. 00			

Acknowledgement

I would like to express my special gratitude to Associate Professor Wang Ke of School of Environment and Natural Resources of Renmin Universiy of China (RUC) for his support on research and survey organizing, data access and sharing of experience and advice. I am extremely grateful for Doctor Dong Changgui of School of Public Administration and Policy of RUC for his contribution on methodology designing and technical support. I would like to acknowledge the active participation of Associate Professor Gong Yanzhen of School of Environment and Natural Resources of RUC. Finally, I am pleased to give my sincere gratitude to Mr. Wang Jijie of National Center for Climate Change Strategy and International Cooperation of China, master students Xiang Qixin, Zhang Wanlin, Xiahou Xinrui, Peng Jing of School of Environment & Natural Resources of RUC, and

Ph D. student Yuan Miao of School of International Studies of RUC, for their assistance in the preparation of the data and original manuscript.

Xu Qinhua

16 Jan 2020

许勤华，中国人民大学国际关系学院教授；中国人民大学国家发展与战略研究院副院长，俄罗斯东欧中亚研究所所长、国际能源战略研究中心主任。为国家能源局研究咨询基地首席专家，中国石油经专委理事，中国能源研究会可再生能专委会秘书长、副主任委员，香港特区政府能源顾问，国务院发展研究中心欧亚社会发展研究所研究员、国家能源局“一带一路”能源合作网专委会首届主任、中国国际文化交流中心“一带一路”绿色发展研究院首任院长、新疆大学客座教授。在国际能源战略、能源国际合作、欧亚地区安全、“一带一路”绿色发展等研究领域发表中英文论文及专业文章百余篇，著作十六部，如《解读中国能源政策：能源革命与“一带一路”倡议》（中英文，2017）、《中国国际能源战略研究》（2014）、*China Energy Policy in National and International Perspectives: A Study Fore-and-Aft 18th National Congress*（2014）、《新地缘政治：中亚能源与中国》（2007）、《低碳时代发展清洁能源国际比较研究》（2013）、《能源外交概论》（2013）、《中国全球能源战略：从能源实力到能源权力》（2017）等，为《中国能源国际合作报告》（2009—2018）主编和中国能源国际合作论坛发起人。主持的中国能源国际合作论坛和“一带一路”圆桌会议及相关成果在国内外相关领域（政府、行业和企业）产生了广泛影响。

Dr. Xu Qinhua is a Professor of International Political Economy and International Relation focused on the strategy and energy at the School of International Studies of Renmin University of China (RUC) . She is the vice dean of National Academy of Development and Strategy, director of Centre for International Energy and Environment Strategy Studies and director of Institute of Russia, Eastern Europe and Central Asia Studies at RUC. She is also the author of 16 books and more than 100 papers and academic articles , such as *China's Energy Policy from National and International Perspectives: The Energy Revolution and The Belt and Road Initiative* (2017), *China Energy Policy in National and International Perspectives: A Study Fore-and-Aft* 18*th National Congress* (2014), *An Introduction of Energy Diplomacy* (2012), *New Geopolitics: Central Asia Energy and China* (2007), *China's Global Energy Strategy: from Energy Strength to Energy Power* (2017), *An International Comparasive Study on Clean Energy Development in Low Carbon Era* (2013), Geographies of Energy Efficiency in China etc.. She holds a number of core positions in various national and international organizations, such as the committee member of WEC (2012 – 2016), senior researcher of APERC (2007 – 2008) . She is now the chief expert at the Research Advisory Base of NEA, China, the standing committee member of Chinese Petroleum Economic Professional Committee, secretary - in - general of Renewable Energy Professional Committee of China Energy Research Society, the reseacher of Development Research Center of the State Council and etc..